How to Construct Your
Intellectual
Pedigree

*A History of
Mentoring in Science*

How to Construct Your Intellectual Pedigree

A History of Mentoring in Science

Elof Axel Carlson

Stony Brook University, New York, USA

World Scientific

NEW JERSEY · LONDON · SINGAPORE · BEIJING · SHANGHAI · HONG KONG · TAIPEI · CHENNAI · TOKYO

Published by

World Scientific Publishing Co. Pte. Ltd.

5 Toh Tuck Link, Singapore 596224

USA office: 27 Warren Street, Suite 401-402, Hackensack, NJ 07601

UK office: 57 Shelton Street, Covent Garden, London WC2H 9HE

Library of Congress Cataloging-in-Publication Data

Names: Carlson, Elof Axel, author.

Title: How to construct your intellectual pedigree : a history of mentoring
   in science / Elof Axel Carlson, Stony Brook University, New York.

Description: New Jersey : World Scientific, [2021] | Includes
   bibliographical references and index.

Identifiers: LCCN 2020034091 | ISBN 9789811215827 (hardcover) | ISBN
   9789811215834 (ebook) | ISBN 9789811215841 (ebook other)

Subjects: LCSH: Scientists. | Intellectuals.

Classification: LCC Q141 .C218 2021 | DDC 509.2/2--dc23

LC record available at https://lccn.loc.gov/2020034091

British Library Cataloguing-in-Publication Data

A catalogue record for this book is available from the British Library.

For any available supplementary material, please visit
https://www.worldscientific.com/worldscibooks/10.1142/11704#t=suppl

*Dedicated to the unbroken line of mentoring
by which knowledge is imparted from
one generation to the next.*

Contents

Preface

I wrote this book so other scientists can construct their own intellectual pedigrees. This makes it a manual and I show that there is no one way to display such a pedigree. This gives the reader lots of choices about the information desired for display as well as the aesthetics of how such a pedigree will look. There are presently several phrases in use for this type of construction: intellectual pedigree, academic pedigree, mentoring pedigree (or fly tree, neurotree, etc.) and all seem reasonable to me. They convey an interest in that process by which knowledge is passed across generations through a variety of ways, the most intimate being a one-on-one relationship between an academic sponsor and a student.

Academics earning a PhD in a scientific research field are usually not trained to become effective teachers. In the life sciences the closest most graduate students come to that experience is their year or years spent as a teaching assistant (often designated by initials alone, "being a TA"). Most freshly minted PhDs learn from their errors or copy the habits of teachers they admired. Most become competent or excel at teaching, but they are cautioned not to overdo their enthusiasm for teaching because that will leave little time for research. The teaching usually at issue is the lecture course especially large courses for undergraduates. While communicating scientific ideas and information is essential for good teaching, there is another aspect that is rarely discussed. Good teachers are good mentors. Mentors are more likely to be associated with one-on-one encounters and nothing has a greater impact in a scientist's training than the mentoring of a graduate sponsor for a dissertation research project. I knew that my mentor, Hermann Muller, spent many hours with me going over my research plans, meticulously editing my write-ups for projects and publishable papers, and encouraging me. He was also demanding in his expectations but never in the sense that I felt I was being exploited. In fact, he made sure his name did not appear on papers that I had conceived and carried out.

I began looking at mentoring relations of scientists by constructing my own intellectual pedigree. I used Sturtevant's 1965 efforts as a guide. But Sturtevant did not provide information on the forerunners of Morgan's legacy. Over the years I read biographical accounts of those in my pedigree and was surprised that I could push this mentoring sequence to notable icons of science like Newton, Darwin, or Galileo. This book reveals a lot of interesting findings and answers some basic questions. How did this mentoring system arise? How did it move from one country to another? How did fields shift in a lineage from mathematics or physics to biology? For those interested in the history, sociology, or philosophy of science, this approach to mentoring is new and what it might lead to is largely unknown. I hope readers will prepare their own intellectual pedigrees and place these on websites. I know little about attempts, if they exist, for intellectual pedigrees in major parts of the world — Chinese, Japanese, Indian, African, Middle East, and Latin American. I used primarily geneticists and traced their mentoring past as far as I could go. While there will be overlaps if this study were done for evolutionists, embryologists, cell biologists, biochemists, and microbiologists, I expect some differences would emerge by field. I included all the geneticists featured in Krishna Dronamraju's book, *A Century of Geneticists: Mutation to Medicine* (2018).

Most of the reading for this I accomplished at home in Bloomington, Indiana. I also used the Wells Library and the Lilly Library at Indiana University for finding information I could not get on the web. I thank Abraham Krikorian and Krishna Dronamraju for their many helpful suggestions.

I thank my daughter Christina Carlson for help preparing the circular diagrams. I thank for helpful suggestions from Krishna Dronamraju, Abraham Krikorian, Shari Cohn, Caitlin Carlson Jones, and many students who supplied information on their careers. For the 36 illustrations in this manual, see the Afterword for information on their sources. All, unless noted otherwise, are from the web and are in public domain because they were published before 1989 in United States publications, or before 1923 (international).

Elof Axel Carlson
Bloomington, Indiana
Emeritus Distinguished Teaching Professor, Stony Brook
University, New York, and
Visiting Scholar and Fellow, Institute for Advanced Study,
Indiana University

1 What are Intellectual Pedigrees?

Intellectual pedigrees are attempts to trace back the influence of scholars on one another, especially through a process of mentoring. Mentors can be teachers or colleagues at a university. They can be family members or grade schoolteachers. Sometimes they can be authors of books or articles that are read during formative years. In my own academic pedigree, I single out one high school teacher and my PhD dissertation advisor as the most influential in my career. I could identify perhaps a dozen with strong influence but not as imposing as these two — **Morris Gabriel Cohen** and **Hermann Joseph Muller**. I applied the same logic to my predecessors.

I begin with the intellectual pedigree of my mentor, Hermann Joseph Muller. Next to myself, he is the scientist about whose life I have the most knowledge. Not only did his mentoring span my graduate years 1953–1958, I had the pleasure of writing his biography — *Genes, Radiation, and Society: The Life and Work of H. J. Muller*. In this intellectual pedigree, I will work back in time to the earliest known of Muller's intellectual predecessors. In each instance I have tried to give a brief one paragraph summing up of an intellectual ancestor or descendent and a portrait, if one were available. This approach connects Muller to a network of scholars most of whom I could not have predicted. Unlike a human genetics pedigree with multiple generations producing progeny, the intellectual pedigree differs in important ways. The chain of

■1

mentoring is not always linear. Some scientists have more than one mentor. This leads to a branching of the intellectual pedigree. Usually the mentor is at the college level. For most of these scientists that experience was in the process of getting an MD or PhD degree.

In the Middle Ages there were four divisions of the University — all took the seven liberal arts for the BA degree. The liberal arts were first described by Plato in *The Republic* but while there were Greek and Roman academies for learning, there were no universities.[1] Plato introduced the liberal arts as the knowledge and thinking required for philosopher-kings in his ideal republic. They were tools for interpreting the "the true, the good, and the beautiful" in his era. Instead of a university with classes, the Greek scholars preferred a single scholar who mentored several students (e.g., Plato, Socrates, and Aristotle had their own schools). The liberal arts were not introduced into the European teaching monasteries until the 6th century by Boethius (477–524 CE) who is sometimes described as "the last of the Romans and the first of the Scholastics". The trivium consisted of grammar, logic, and rhetoric. The quadrivium consisted of arithmetic, geometry, music, and astronomy. Students could then choose one of three graduate ("higher faculty") specialties — law, medicine, or theology, for a MA, MD, or PhD level of knowledge. In all these specialties, students mastered what was known in these fields. It was not until 1810 that the German PhD became a research dissertation degree in which students did original research and defended their theses (Master's degree) or dissertations (PhD degree) in front of their faculty.

The university is an invention of European origin.[2] It was a guild-like arrangement of masters and scholars with two traditions emerging. One had the students hiring the masters and running the university. The more widespread organization was a faculty of masters who ran the university and charged the students fees for entry. Both were affiliated with the Catholic Church. The Medieval Universities differed from the Madrasas that followed the expansion of Islam from the Middle East to Spain. The Madrasas were focused on the Koran and its commentaries. They were not separate entities established by scholars or students. They were affiliated with a mosque and did not initially have formal academic degrees as did the Medieval Christian Universities. Most Islamic scientists were supported by patrons or their rulers. The University of Bologna in Italy, in 1088, was the first university established in Europe (and the students ran it). The University of Paris was established in 1150 (and the faculty ran it). The first public supported university was the University of Naples in 1224. By 1413 there were 21 universities in Europe. The first university in North America was Harvard University (1636).

In the 19th century both the MD and the PhD required a dissertation of book length. The research and dissertation aspects were dropped from the MD in the 20th century. The PhD was not part of the English academic degree. Until the twentieth century the British MA was the terminal degree of higher education. It was the flow of American students to German Universities that motivated Great Britain in 1917 to introduce the PhD so that American loyalties would be for Britain rather than Germany if another war broke out between Germany and Great Britain. The modern university PhD was established by the **von Humboldt** brothers, **Alexander** (1769–1859) and **Wilhelm** (1767–1835).[3] There was (and still is) a degree called the DSc (Doctor of Science) used in Britain. It was awarded by a faculty committee to a scholar with a proven record of recognized research. It was rare and like election as a Fellow of the American Association for the Advancement of Science. **Alexander** and **Wilhelm von Humboldt** believed scientists should contribute to new knowledge by their experiments and discoveries and the research dissertation was the vehicle for launching new scientists and new knowledge. Another feature of higher education before the 19th century was the religious nature of most universities. They were highly motivated to produce ministers or priests. There were no secular universities in the Middle Ages or the Renaissance. Many scientists of those more distant centuries saw their careers as priests interpreting the works of God. Women were effectively absent from higher education until the late 19th century. The first female professor was **Laura Bassi** (1711–1778) who was born in Bologna and got her degree at the University of Bologna in 1732. She was appointed Professor of Natural Philosophy (Physics) and helped spread the work of Newton to her students. She did research in chemistry, physics, mathematics, hydraulics, and mechanics.

In the United States, after the Revolutionary War, most colleges and universities were undergraduate institutions primarily training males for the ministry or, as phrased in their mission statements "creating Christian gentlemen." American scientists took their BA degree and then went to Germany for a PhD. The first American PhD was awarded by Yale in 1861. In the 1870s the University of Pennsylvania, Harvard, Princeton, Clark, and Johns Hopkins established PhD programs based on the German model of research. **Daniel Coit Gilman** at Johns Hopkins University played a major role in promoting the German PhD model in American Universities.[4] The two major suppliers of PhDs in the life sciences were Harvard University and Johns Hopkins University. Harvard University did so by recruiting Swiss scientist **Louis Agassiz** from Paris to its faculty. Johns Hopkins University staffed its faculty with **H. Newell Martin** (a student of **Thomas Henry Huxley**) and **William Keith Brooks**, a student of Agassiz.

Recommended Reading

1. Plato (375 B.C.) *The Republic.* The Jowett translation is available free on Project Gutenberg.

2. Rashdall, Hastings (1858–1920) *The Universities of Europe in the Middle Ages* (1895) Cambridge University Press) covers the history of the major European universities. It is a large work (1500 pages) and now available as an eBook.

3. For an account of the history of the university after the Middle Ages and the Renaissance, see Anderson, Robert D. *European Universities from the Enlightenment to 1914.* It includes a lengthy discussion (Chapter 4) of the Humboldt brothers and their establishment of the University of Berlin in 1810. The idea of a university as a place for academic freedom to pursue scholarship and original research was shaped by the Humboldt brothers.

4. For a biographical account of the contributions of Gilman, see Franklin, Fabian *The Life of Daniel Coit Gilman* (1910 NY Dodd, Mead and Company).

2 The Origin of Academic Pedigrees

People have used pedigrees for many endeavors. The idea goes back to Biblical times and the Old Testament has numerous lists of family members across dozens of generations. They were first used in a more structured way to identify family histories or genealogy. This included royal families. Present monarchs in countries with a royal lineage can trace these pedigrees back to the Middle Ages or earlier. In the mid-Nineteenth Century **Francis Galton** used pedigrees to study talents and other traits running in families (Figure 1). His own family of Galtons, Darwins, and Wedgwoods is an example [1]. Galton's pedigree analysis of intellectual and sports ability were forerunners of the pedigrees used in the eugenics movement especially by **Charles Davenport** and **Harry Laughlin** at the Eugenics Record Office in Cold Spring Harbor, NY [2]. Many geneticists recognized that these were highly influenced by bias when they included traits like becoming bankers, carpenters, physicians, or sea captains, among the talented and becoming criminals, paupers, or mentally deficient (in those days the terms "feebleminded" and "unfit person" were used) [3]. After WWII and the abuses of Nazi racial ideology that expanded the approach started by the Eugenics Record Office, such pedigrees for social traits were avoided. Instead they were used for taking family histories by physicians and studied in detail for medically significant traits for a field that became known as genetic counselling pioneered by **Sheldon Reed** (1910–2003) [4]. An additional use of

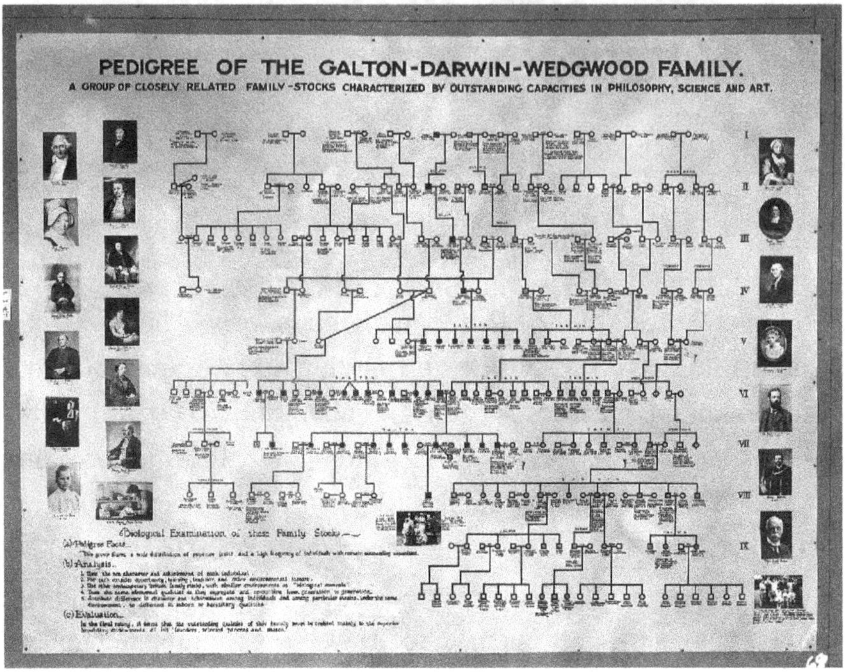

Figure 1:   The Galton-Darwin-Wedgwood pedigree used by the Eugenics Record Office at Cold Spring Harbor to illustrate inheritance of talent and high intelligence.

family pedigrees was introduced by **Murray Bowen** (1913–1990) to work out family dynamics among relatives across three generations. A religious use of pedigrees is widely used by the Latter Day Saints (LDS or Mormons) who require members to prepare extensive pedigrees of their ancestral families so that they can be assimilated into the Mormon family [5].

Galton used dictionaries of biography and scores obtained from competitive examination among top students in mathematics, classics, and other fields of knowledge. These pedigrees were also used in the eugenics movement that Galton founded, especially by **Charles Davenport** and his staff at the Eugenics Record Office in Cold Spring Harbor during the first third of the twentieth

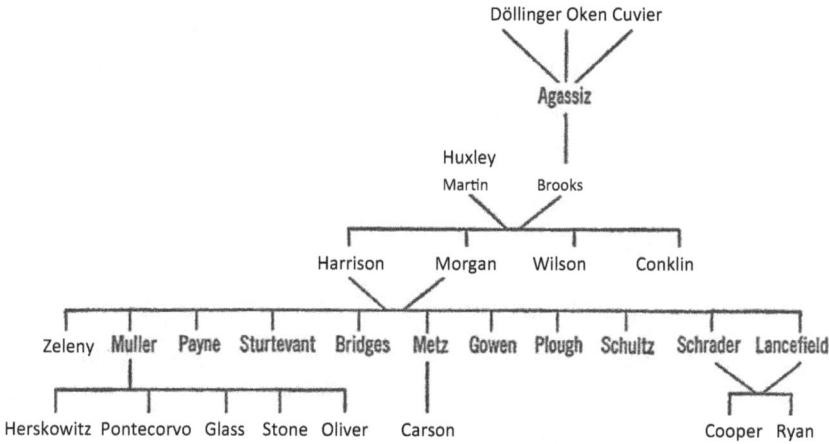

Figure 2: Sturtevant's intellectual pedigree showing Morgan's mentors to Döllinger and Morgan's 11 students and 8 of their students. Irwin Herskowitz was Muller's laboratory coordinator. His PhD was with Dobzhansky.

century. Earlier pedigrees were largely used for the nobility who traced their ancestry to the founding dynasties of their present rulers. The first academic pedigrees to appear as mentoring relations among geneticists were introduced by **Alfred Henry Sturtevant** in his 1965 book, *A History of Genetics* [Figure 2] [5]. Today, there are several sources for scholars to trace their mentors. The Academic Family Tree is probably the largest with 38 disciplines on file. There are separate files on-line for physicists, mathematicians, philosophers, neurologists, and chemists. There is also an automatic program that can generate pedigrees of PhD mentors going back at least to the first dissertation-based academic PhDs that began under the Humboldt brothers, Alexander and Wilhelm, in the early nineteenth century.

Intellectual pedigrees can be a simple display of names connected by lines or they can be accompanied by a photo of the scientist's face. I have used a down pointing arrow to indicate the student to mentor direction (↓). I enriched the value of the intellectual academic pedigree for **Muller** (and myself) by including a brief account of the major contributions of each distant mentor. A larger number, 63, of linear representations of geneticists (with briefer biographies) is provided in the Appendix.

Notes and References

1. Galton Francis (1869) *Hereditary Genius*. MacMillan, London.
2. Carlson Elof Axel (2001) *The Unfit: A History of a Bad Idea*. Cold Spring Harbor Laboratory Press, New York. Chapters 13 and 15 describe the use of pedigrees for claiming a hereditary basis of social traits and for advocating compulsory sterilization laws to prevent the unfit from reproducing.
3. Muller H J (1932) The dominance of economics over eugenics. *Third International Congress of Eugenics, New York A Decade of Progress in Eugenics* Williams & Wilkins, Baltimore, Maryland.
4. Reed Sheldon (1955) *Counseling for Medical Genetics*. Saunders, Philadelphia.
5. Bowen Murray (1985) Therapy *in Family Practice*. Rowman and Littlefield Publishers, Lanham, Maryland.
6. Sturtevant Alfred (1965) A *History of Genetics*. Harper and Row, New York.

3 How to Prepare Your Academic Pedigree

It is easiest to do this in stages. The first stage is getting the chronological sequence. Place the name of each person, the years of birth and death and places where they were born and died.

Thus: Hermann Joseph Muller (1890–1967)
        b. New York City, NY; d. Indianapolis, IN.

Most of the names will be available from Wikipedia when you use a computer search engine like Google. Most of the Wikipedia entries for scientists have a box in the upper right with a photograph and a summary of significant facts (birth, death, parents, education, mentor, and students). The most significant mentor in a scientist's life is usually his or her mentor for a PhD or MD. The information you are seeking has a high probability of being found in Wikipedia. If not, check the other options the Google search list provides. This may be obituaries, retirement tributes, or encyclopedia entries (especially from the 11th edition of the *Encyclopedia Britannica*). For American scientists, there is usually an extensive obituary from the *Biographical Memoirs of the National Academy of Sciences* which is free to access on the web and read. For information on older scientists, I recommend using the *Dictionary of Scientific Biography*.[1]

I take notes for the page devoted to a scientist I wish to follow. This includes aspects of his or her career, significant contributions to science, and interesting items about coping with reverses, injuries, political or religious clashes, or little-known facts. Often key persons other than the PhD mentor are mentioned and these can be branches of the pedigree for that scientist.

After amassing the facts from reading the Google search, I choose a suitable portrait (if there is one). I recommend you do it yourself (or with your academic mentor) and that way you can quickly find what you want in your first search.

Writing My Own Entry

In this entry I use the first person. When Mendel applied for a position as a teacher, he submitted his biography in the third person. If I were preparing this for a student I had mentored, I would use the third person. I would also use the third person if I wanted to hang a copy on the wall of my office or study. If I were giving this pedigree to my family members, I would use the first person for my own entry. Most of the 60 or so pedigrees I have done involve about 15 to 20 entries to go back to the Renaissance. These can be entered in a notebook on one page and later be put on to 3×5 index cards with notes on the person's life.

First-person Narrative for Elof Axel Carlson

Elof Axel Carlson — I am a geneticist and historian of science with a love for teaching and scholarship. I was born in Brooklyn, NY on July 15, 1931. My father was from Stockholm, Sweden. My mother was from Bound Brook, New Jersey. As a teacher I am aware of the rippling effects of lectures, conversations during office visits, and discussions both casual and formal in laboratories. In my own life I was influenced by my father, **Axel Elof Carlson** an elevator operator who had a passion for reading and whose library was a source of constant surprises as I browsed through the books growing up. My mother was the first child of immigrant parents from Ternopil, in present day Ukraine. I was influenced by many of my school teachers, especially in Grades 7 to 12 (junior high school and high school). The most significant of these teachers was **Morris Gabriel Cohen** at Thomas Jefferson High School in Brooklyn, NY. I read aloud to him from the classics over a period of five years, meeting him for an hour about 7 a.m. five days a week while school was in session. I attended NYU on a scholarship, and majored in biology and minored in history. I was accepted to Indiana University in Bloomington, Indiana, where I studied genetics with Nobelist **H. J. Muller**. I have had the pleasure of supervising the PhD dissertations of six students (and seeing 13 of my books published). My most noted books are *The Gene: A Critical History* (1966), *Genes, Radiation, and Society: The Life and Work of H. J. Muller* (1981), *The Unfit: A History of a Bad Idea* (2001), and *Mendel's Legacy: The Origins of Classical Genetics* (2007). My PhD was on the structure and mutability of the dumpy gene in fruit flies. My laboratory research involved comparative genetics, gene structure, and mosaicism. I applied my work on mosaicism to medical genetics using retinoblastoma as an example. I have published articles on gene structure, chemical mutagenesis, and human genetics. I taught at Queen's University in Canada (Queen's University), at UCLA, and at Stony Brook University before retiring with my wife, Nedra, in Bloomington, Indiana.

Third-person Narrative

Elof Axel Carlson — is a geneticist and historian of science with interests in teaching and scholarship. He was born in Brooklyn, NY, on July 15, 1931. His father was from Stockholm, Sweden; and his mother was from Bound Brook, New Jersey. As a teacher, he was aware of the influences of rippling effects of lectures, conversations during office visits, and discussions both casual and formal in laboratories. He was influenced by his father, **Axel Elof Carlson**, an elevator operator who had a passion for reading and whose library was a source of constant surprises as his two children browsed through the books growing up. His mother was the first child of immigrant parents from Ternopil, in present day Ukraine. He was influenced by many of his school teachers, especially in Grades 7 to 12 (junior high school and high school). The most significant of these teachers was **Morris Gabriel Cohen** (1900–1975) at Thomas Jefferson HS in Brooklyn, NY. Carlson read aloud to him from the classics over a period of five years, meeting him for an hour about 7 a.m. five days a week while school was in session. Carlson attended NYU on a scholarship, and majored in biology and minored in history. He was accepted to Indiana University in Bloomington, Indiana, where he studied genetics with Nobelist **H. J. Muller**. He supervised the PhD dissertations of six students (and seeing 13 of his books published). His most noted books are *The Gene: A Critical History* (1966), *Genes, Radiation, and Society: The Life and Work of H. J. Muller* (1981), *The Unfit: A History of a Bad Idea* (2001), and *Mendel's Legacy: The Origins of Classical Genetics* (2007). Carlson's PhD was on the structure and mutability of the dumpy gene in fruit flies. His laboratory research involved comparative genetics, gene structure, and mosaicism. He applied his work on mosaicism to medical genetics using retinoblastoma as an example. He published articles on gene structure, chemical mutagenesis, and human genetics. He taught at Queen's University in Canada, at UCLA, and at Stony Brook University before retiring with his wife, Nedra, in Bloomington, Indiana.

Comment on First- or Third-person Usage

Note that the first-person account gives a more intimate feeling about the writer although none of the facts have changed. Third-person usage objectifies the writer. First-person accounts are usually associated with informal correspondence. A major use of first-person accounts are applications for admission to college or graduate programs or application to admission to medical school. Medicine is both an art and science. Good medical practitioners need to relate to patients and that helps patients follow the health advice given by physicians. The first-person narrative often conveys this empathy. The CV or *curriculum vitae* is usually written in the third person and is used for job hiring and promotions or when applying for grants. It is usually structured when there are competitive positions and thus the term "fill out a CV" is more often used than "compose a CV." If it is composed, it is more often associated with activities like submitting a book manuscript or prospectus for publication.

In my own academic pedigree, I have chosen to include my high school mentor, **Morris Gabriel Cohen** because of the profound influence he had on my early education. I had many other mentors as an undergraduate at NYU and as a graduate student at Indiana University. This is probably true for most of the persons whose pedigrees are described. If all mentoring were included, the pedigree would resemble a bush more than a tree. Thus, I chose to use Morgan's academic mentors at Johns Hopkins, H. Newell

Martin and William Keith Brooks. I did not include his European mentors when he went abroad for a year after getting his PhD. There he formed a lifelong friendship with **Hans Driesch** and absorbed much of the German advances in experimental embryology. He also found a visit to **Hugo de Vries'** garden a stimulus for starting his own studies of mutation.

Note

1. Gillispie Charles C. (editor) (1970–1980) *Dictionary of Scientific Biography*, 16 volumes. Charles Scribner's Sons, New York.

4 First Mentoring Influence on Elof Carlson — Morris Gabriel Cohen

Morris Gabriel Cohen (1900–1972) was a teacher at Thomas Jefferson High School in Brooklyn, NY. His father was a cigar maker and a friend of Samuel Gompers, the founder of the American Federation of Labor. Cohen was drafted in WWI and served overseas in England and France. He returned to New York and attended Columbia University on a scholarship. He said he "majored in everything useless".

He participated in the first presentation of Columbia's famous Contemporary Civilization of the West. While in college his vision deteriorated, and he was diagnosed with Leber's optic atrophy, a progressive loss of vision whose cause was unknown at the time. It is a maternally inherited mitochondrial disease with retinal degeneration. He dropped plans for graduate school and became a high school teacher. He taught history, but I knew him through a lucky occurrence. I was sitting in the dark in an empty auditorium using an exit light to read.

He walked in and must have heard me moving. He asked me what I was reading and why I was reading there. I told him I liked to read, and I came in early because my father had to leave early to get to his job and I left shortly after he did. He invited me to his office and asked me if I would like to read aloud to him. We started with the plays of Aeschylus, Sophocles, and Euripides and then went to the trio of Socrates' arrest, trial, and death and worked our way over the next five years (the last four while I was at NYU) ending with Freud's *Civilization and its Discontents*. It was the most transforming intellectual experience of my life to read these aloud and discuss their implications for our lives. I did not know it at the time, but I was experiencing the mentoring of a millionaire's child in the 1700s when private tutors shaped the minds of the children of the wealthy. He encouraged students to get PhDs. He felt law and medicine were well represented by talented students and that PhDs sustained and added to the richness of civilization.

5 Second Mentoring Influence on Elof Carlson: Hermann Joseph Muller

Hermann Joseph Muller (or H. J. Muller for almost all his publications) was born in NYC in 1890 and died in Indianapolis in 1967. He received a scholarship and attended Columbia University for his undergraduate work in zoology. His most influential teacher during those undergraduate years was **Edmund Beecher Wilson**. He then got an MA at Cornell University and in 1912 returned to Columbia to work with **Thomas Hunt Morgan**. Muller was a founding member of the Fly Lab as Morgan's group was called. They worked out many of the features of classical genetics. Muller's PhD was on a phenomenon called coincidence and interference which was used to prepare accurate maps of genes on a chromosome. He taught at Rice Institute and then at the University of Texas where his most famous work, demonstrating that X-rays induced gene mutations took place. He helped establish the field of radiation genetics and worked on gene size and number, the functional classification of genes, and the time of origin of gene mutations. For this work he received the Nobel Prize in Physiology or Medicine in 1946.

He was controversial in his efforts to promote radiation protection especially during the Cold War. He advocated positive eugenics but denounced the negative eugenics movement as racist, sexist, and based on false assumptions on how social status arises. Muller also taught in the USSR, at Edinburgh, at Amherst College, and at Indiana University. **Elof Carlson** was Muller's student from 1953 to 1958. Among Muller's other MA, PhD, and postdoctoral students are: **Bentley Glass, Clarence P. Oliver, Wilson S. Stone, George Snell, Raissa Berg, Guido Pontecorvo, Charlotte Auerbach, Alexandra Prokofeyeva, S. P. Ray Chaudhuri, Daniel Raffel, Carlos Offermann, Irwin Oster, Abraham Schalet, Seymour Abrahamson, Wolfram Ostertag, Dale Wagoner, Shanta Iyengar, Rafael Falk, William Trout III,** and **Sara Frye**. He also mentored **Carl Sagan** who spent a summer in Muller's laboratory.

6 Muller's Intellectual Pedigree Through Morgan

Thomas Hunt Morgan (1866–1945) was born in Lexington, Kentucky, to a famous family (his father and brother led Morgan's Raiders during the Civil War) and his maternal grandfather was Francis Scott Key. Morgan downplayed his genealogy and treated his students as colleagues. He died in Pasadena, California. He got his PhD in 1890 with **William Keith Brooks** but was more stimulated in his interests with invertebrate development by physiologist **H. Newell Martin**. Morgan's PhD was on the embryology of sea spiders (picnogonids). He went to Naples and was influenced by the new German school of developmental mechanics (Entwicklungsmechanik). Morgan taught at Bryn Mawr and then joined the faculty at Columbia University where he founded the Fly Lab, using Drosophila to find X-linked inheritance and a phenomenon of crossing over within paired homologous chromosomes that accounted for the considerable recombination of traits during sperm or egg production. Morgan received the Nobel Prize in 1934 for his work. He moved to CalTech in 1928 and established it as a world class center for genetic research. Morgan made substantial contributions to embryology before he switched to genetics. He studied the regeneration of limbs and the production of twins and chimeras by experimental means. His switch to genetics came from his association with **E. B. Wilson** and his visit to **Hugo DeVries** in Holland who claimed he observed new species arising in fields of primroses (*Oenothera lamarckiana*). In addition to Muller, Morgan's PhD and postdoctoral students included **Fernandus Payne, A. H. Sturtevant, C. B. Bridges, C. C. Tan, Jack Schultz, George Beadle,** and **Otto Mohr.**

7 Muller's Mentor Lineage to Morgan Via H. Newell Martin

Note: **Morgan** had two mentors at Johns Hopkins, **Henry Newell Martin** and **William Keith Brooks**. The intellectual lineage of Martin leads to **Charles Darwin and Isaac Newton**. The intellectual lineage from Brooks leads to **Galileo** and the fifteenth century. After the discussion of Morgan to Darwin through Martin, we will shift to Brooks' intellectual lineage which goes back to 1499. Thus, there are two branches that fed into Morgan at Johns Hopkins University.

H. Newell Martin (1848–1896) was born in Newry, Ireland, and died in London, England. His father was a Congregational minister. He was educated at University College in London and assisted **Thomas Henry Huxley**. He studied physiology with **Michael Foster** and made it his life's work. He was especially interested in the effects of temperature and exercise on heart beats. Cardiac physiology was Foster's specialty. In 1876 he was hired by President **Daniel Coit Gilman** for the new university named for Johns Hopkins on recommendation from **William Keith Brooks**. At Johns Hopkins he introduced students to experimental methods and **Thomas Hunt Morgan**

viewed his approach as more satisfactory than that of Brooks, who was his sponsor. Morgan found Brooks too philosophical in approach. Morgan, however, felt that the problems that Brooks championed, a search for experimental ways to study heredity and variation, were of far greater significance than those worked on by Martin. Martin married the widow of a Confederate General. He returned to England, however, because his wife died in 1892 and his health was impaired from alcoholism. He died young at 48.

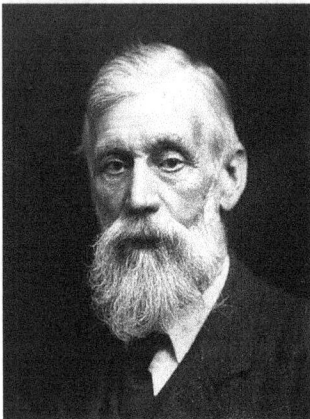

Michael Foster (1836–1907) was born in Huntingdon, England. His father was a physician. Foster attended University College, London, where he studied with **Thomas Henry Huxley**. He became a physiologist and his two most famous students were **H. Newell Martin** and **Charles Scott Sherrington**. **Sherrington** won the Nobel Prize in 1932 for his work demonstrating synaptic communication among neurons in the brain and in muscle cells.

Thomas Henry Huxley (1825–1895) was born in Ealing, England, and died in Eastbourne, England. His father was a mathematics teacher and when the school failed, young Huxley had to drop out of his father's school at the age of 10. He became an autodidact and taught himself Latin, Greek, and German and read widely. He served as an apprentice to surgeons and joined the Navy as an assistant surgeon. He was sent on the *HMS Rattlesnake* to an expedition to New Guinea and Australia. From his dredging at sea he collected numerous medusae and classified them. His surgical skills were exceptional, and he found these organisms had only two layers instead of three (they lacked mesoderm). His papers on what he called the Hydrozoa (today called Cnidarians) earned him election to the Royal Society at the age of 26. Huxley became friends with **Charles Darwin** and while he was initially skeptical of evolution and natural selection (which Darwin confided to him), he was won over by the evidence and became a strong supporter when Darwin's theory came out. Huxley taught at the London School of Mines and there he developed a theory of liberal arts education which was influential in Europe and North America. His essay "A liberal education and how to get it" was published in 1868. His essay "On a piece of chalk" is one of the most captivating public lectures ever given on the significance of evolution. Huxley rejected formal religious belief and coined the term agnostic to represent his views that if a belief lacks scientific demonstration it can neither be proven nor disproven. His major students were **Henry Newell Martin** and **Michael Foster**.

Charles Darwin (1809–1882) was born in Shrewsbury, England, to a wealthy family. He died at his home, in Downe, England. His father was a physician and his mother was from the Wedgwood family of the pottery industry. His grandfather, **Erasmus Darwin**, was a member of the Lunar Society that brought the Enlightenment to England. Charles Darwin was sent to Edinburgh to learn medicine but did not like it. He transferred to Cambridge to study for the ministry and there enjoyed geology with **Adam Sedgwick** and botany with **John Stevens Henslow**. Henslow recommended Charles Darwin for the position of ship's naturalist for the voyage of the *HMS Beagle* (1831–1836). Darwin's account of the round the world trip established him as an appealing writer and a scientist of first rank. His letters to Henslow while on the trip were published and thus Darwin found himself well known to scientists on his return. Darwin developed a theory of evolution. First, he amassed evidence that evolution occurred using the fossils he found and his analysis of the distribution of life in the places he explored. Second, he proposed a theory of natural selection in which life is modified each generation by the survival or loss of competing individuals within and among species. He did not publish his theory until 1858. In 1859 a summary of some 300 pages was published as the *Origin of species*. It made Darwin one of the most controversial scientists of all time. It also transformed the life sciences. Darwin did not teach at universities. He used his inherited wealth to become a full-time scholar and he wrote at his home near London. He contributed to the careers of **Joseph Hooker** and **Thomas Henry Huxley**. He was influenced by the writings of **Thomas Malthus** and **Charles Lyell**. Malthus's essay on population gave him the idea of natural selection. Lyell's geology suggested a relatively similar climate and geological processes occurring over long periods of time, a theory called uniformitarianism.

Adam Sedgwick (1785–1873) was a British geologist. He was born in Dent, England, and died in Cambridge, England. He was educated at Cambridge and taught geology there. He identified the Devonian era as one that followed Cambrian rock formation and believed there was a series of floods followed by new creations by God. In 1831 he invited Darwin to do a geological study of Scottish rock formations when Darwin was a student at Cambridge. Although they remained friends, Sedgwick vehemently rejected Darwin's evidence of evolution and its mechanism of natural selection.

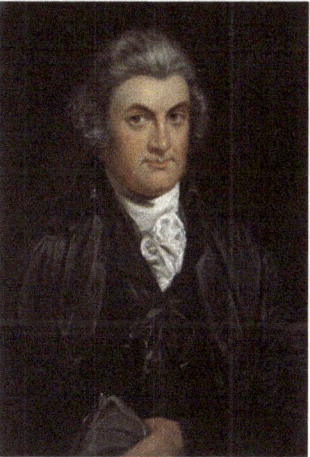

Thomas Jones (1756–1807) was a mathematician and gifted teacher. He taught at Cambridge. He mentored **Adam Sedgwick**.

Thomas Postlethwaite (1731–1798) was a mathematician at Cambridge who mentored **Thomas Jones**. Postlethwaite was mentored by **Stephen Whisson**.

Stephen Whisson (1710–1783) was a mathematician at Cambridge who taught mathematics for physicists. He mentored **Thomas Postlethwaite**. He was mentored by **Walter Taylor**.

Walter Taylor (1700–1743) (no photo available) taught mathematics and Greek. He taught **Stephen Whisson**. He was mentored by **Robert Smith**.

Robert Smith (1689–1768) taught at Cambridge and extended Newton's laws to astronomy. He mentored **Walter Taylor** and he was mentored by **Roger Cotes**.

Roger Cotes (1682–1716) was born in Burbage, England, and died in Cambridge, England. He was Newton's student. He helped expand Newton's second edition of the *Principia* and worked on logarithms, especially the spiral found in snails and other living things. He died of a fever at the age of 33, much lamented by **Newton** who felt a brilliant career was cut short.

Isaac Newton (1642–1726) was born in Woolsthorpe, England, and died in Kensington, England. He is one of the greatest scientists and mathematicians. He went to Cambridge and stood out for his mathematical gifts. His teacher, Isaac Barrow resigned as Lucasian Professor and gave the title to Newton. His *Principia Mathematica* led to a view of science as deterministic, governed by laws of nature that could be described mathematically. It led to the "clockwork universe" of the Enlightenment. He introduced calculus as a branch of mathematics (co-discovered by **Leibniz**). Newton was equally committed to studying God and hoped his ventures into biblical scholarship would lead him to understand God's works. He rejected the Trinity and adopted a Unitarian view of Christianity but did not publish his views because they were heretical. He became Keeper of the Mint (equivalent of being Secretary of the Treasury) and made a fortune in his investments. He never married and may have died a virgin.

8 Muller's Lineage from Morgan's Second Mentor, William Keith Brooks

H. J. Muller's undergraduate years were inspired by the courses taught by **Edmund Beecher Wilson**. His graduate years were spent, however, in the laboratory of **Thomas Hunt Morgan**. Both Wilson and Morgan received their PhDs from the same mentor, **William Keith Brooks**. Hence Wilson's pedigree follows the same lineage through Brooks as does Morgan.

Edmund Beecher Wilson (1856–1939) was born in Geneva, Illinois, and died in New York City, NY. In his youth he considered becoming a cellist. His father was a judge. Instead, Wilson went to Yale and loved learning about nature. He went to Johns Hopkins for his PhD studying with **William Keith Brooks**. He chose embryology and cell biology as his special interest and taught first at Bryn Mawr and then at Columbia University. As chair of that department, he recruited **T. H. Morgan** who came to Johns Hopkins several years after Wilson had graduated with his PhD. In 1896 Wilson published the first edition of *The Cell in Development and Inheritance*. It summed up the latest findings in cell biology and Wilson predicted that the nucleic acids might be the source of the hereditary component of the chromosomes. His laboratory work focused on the chromosomes and he used beetles to work out the sex chromosomes, which he designated as X and Y with XX as female and XY as male (with some species having a single X male and two X females and no Y chromosome which he designated as XO male and XX female). Independently, **Nettie Stevens** found sex chromosomes in Diptera that she studied. His student **Walter Sutton** was the first to apply Mendelian inheritance to meiosis and with Wilson and **Theodor Boveri** they called this the chromosome theory of heredity.

William Keith Brooks (1848–1908) was born in Cleveland and died in Baltimore. He had frail health from a congenital heart defect. He was an avid scholar and read widely in classics and philosophy before committing himself to natural history after reading Darwin's *Origin of Species*. At Harvard he got his PhD with **Louis** and **Alexander Agassiz**. He studied embryology of tunicates (the genus *Salpa*) and invertebrates, especially mollusks. He believed a study of tunicates would contribute to the understanding of the evolution of vertebrates from their invertebrate ancestors. He was a gifted teacher and among his PhD and postdoctoral students were **William Bateson, T. H. Morgan, Edmund B. Wilson, H. V. Wilson,** and **Ross Harrison**. He wrote seven influential books, including *The Law of Heredity* (1883). His views were Lamarckian. He told Bateson that the most interesting area he could enter was heredity. He instructed his students to work with living specimens and study their living functions. Bateson followed that advice but rejected Brook's theoretical approach.

Louis Agassiz (1807–1873) was born in Môtier, Switzerland, and died in Boston, Massachusetts. He studied medicine at Zurich, Heidelberg, and Munich and he studied natural history at Erlanger, Paris, and Munich. He did his research with **Alexander Cuvier** and **Alexander von Humboldt**. Cuvier inspired him to become an ichthyologist and Humboldt convinced him of the importance of geology for a study of natural history. Agassiz gained fame by descending through a vertical tunnel he dug into a glacier to examine if water flowed under it. He wrote numerous volumes on fresh water fish and fossil fish. From their distribution he concluded that Europe was covered by glaciers that he described as the Ice Age. When he visited the United States to study North American fish, he fell in love with American society and accepted an appointment at Harvard. His students included **David Starr Jordan** and **William Keith Brooks**. He established a forerunner of Woods Hole laboratories at Penikese Island in Buzzard's Bay, inspiring Brooks to set up a similar, but portable, summer laboratory by the sea in Beaufort, Maryland. He opposed natural selection but never criticized his students who were all Darwinians. At Woods Hole his motto greets visitors: "Study Nature, Not Books."

Georges Cuvier (1769–1832) was born in Montbéliard, and died in Paris, both in France. He was raised as a Lutheran, a faith he kept throughout his life. His father was in the Swiss Guard. Young Cuvier had a near photographic memory and remembered what he read in prodigious amounts including Buffon's volumes on natural history. He studied at Stuttgart and then in Paris where he gained attention for his studies of living and fossil elephants. He described Asian and African elephants as separate species and the fossil elephant (that he named mastodon) also a separate species. He received his doctorate from **Ignaz Döllinger**. Cuvier is a founder of the field of comparative anatomy and made numerous contributions to taxonomy, the use of stratigraphy to represent layers of animals that became fossilized. He explained this with a theory of sudden extinctions (attributed to massive tsunamis) and replacements with new creations. He opposed vehemently Lamarck's theory of evolution and modification by use and disuse and wrote a scathing eulogy in 1832 denouncing Lamarck's character. Cuvier's most famous student was **Louis Agassiz**.

Ignaz Döllinger (1770–1841) was born in Bamberg, Germany, where his father was a Professor and physician. He died in Munich. He got his doctorate in 1794 and studied in Würzburg, Padua, and Vienna before settling in to Munich where he was Professor of Anatomy. His medical degree was from Padua and his doctoral advisor was **Antonio Scarpa**. Döllinger taught medicine as a natural science. He devoted most of his research to embryonic development. His students included **Louis Agassiz, Karl Ernst von Baer, Christian Pander, Lorenz Oken**, and **Johann Lukas Schönlein**. Pander and von Baer helped establish the field of embryology. **Pander** introduced the idea of three germ layers in the early embryo — ectoderm, mesoderm, and endoderm. It was **von Baer** who described development as epigenetic with organs formed not by enlargement but by differentiation of embryonic germ layers. **Schönlein** was one of the first German Professors to lecture in German rather than in Latin. **Oken** was a founder of a holistic, transcendental approach to biology called *naturphilosophie* that was extended by **Goethe**. It tried to make sense of homologous structures and how they came to be modified.

Antonio Scarpa (1752–1832) was born in Liorenzaga, Italy, and died in Padua, Italy. He received his medical degree at the University of Padua, studying with **Giovanni Battista Morgani**. He studied the anatomy of the inner ear and cardiac nerves. He taught at the University of Modena. He was a bachelor and had several children out of wedlock. Born poor, he became wealthy, and a collector of art. Scarpa's most noted student was **Ignaz Döllinger**.

Giovanni Battista Morgagni (1682–1771) was born in Foril, Italy, and died in Padua, Italy. He was a professor of anatomical pathology at the University of Padua. He rejected the prevailing view of the time that disease was systemic (influenced by vital humors or toxins). Instead he argued that it had a localized origin and was usually organ specific. He wrote a five-volume treatise, *On the Seat and Causes of Diseases* based on 640 dissections he carried out. His most famous student was **Antonio Scarpa.** His mentor was **Antonio Maria Valsalva**.

Antonio Maria Valsalva (1666–1723) was born in Imola, Italy. He studied the anatomy of the throat and named the Eustachian tube. His mentor was **Marcello Malpighi**.

Marcello Malpighi (1628–1694) was born near Bologna, Italy. He has been described as "the father of microscopic anatomy, histology, physiology, and embryology." He discovered the trachea network in insects and associated them with breathing. He discovered the Malpighian tubes in kidneys and demonstrated that the pigment of Africans was in a lower layer of the skin. He identified the capillaries as the structures that connected arterial blood to venous blood flow. His colleague **Giovanni Alfonso Borelli** at Pisa introduced him to experimental science and Malpighi chose the microscope for his studies. He was an excellent artist and drew careful illustrations of his microscopic specimens. He demonstrated that galls found in plants were caused by insects that laid eggs in the plant tissue. The Royal Society in London published many of his findings in plant and animal cellular anatomy.

Giovanni Alfonso Borelli (1608–1679) was born in Naples of a Spanish father and Italian mother. He took an interest in experimental physics and mathematics but became a physician. For his medical studies, he took an interest in animal motion and worked out the relation of muscles and bones to limb motion. He was also the first to note the existence of stomata in plant leaves. He is considered the founder of the field of biophysics. He was mentored by **Benedetto Castelli** with whom he studied the detailed orbits of Jupiter's moons.

Benedetto Castelli (1578–1643) was born in Brescia, Italy, and became a mathematician and taught at the University of Padua before becoming the Abbott of Monte Casino. He worked on sunspots with his mentor **Galileo Galilei.**

Galileo Galilei (1564–1642) was born in Pisa, Italy. His father was a composer and played the lute. He wanted Galileo to become a physician but while in medical school Galileo found courses on astronomy and science more interesting and convinced his father to let him change his field. He studied mathematics and made contributions to the physics of falling bodies, using experiments to demonstrate these laws. He moved to the University of Padua where he taught and wrote most of his books. He also made a telescope and applied it to the skies. He discovered the moon had craters, Venus had phases like the moon, Jupiter had four moons, and the sun had sunspots that followed its rotation. He also described Saturn as having "ears" because the rings at that time were tilted and the lenses were not as sharp as later lenses that revealed these were rings. He felt he had the evidence for the Copernican theory and began a series of disputes with his fellow astronomers. He replied in polemic style ridiculing his opponents, many of them Jesuits or influential in the Church. This led to his eventual trial and conviction as a heretic forcing recantation and household arrest for the rest of his life. He is considered one of the greatest scientists of all time and he helped launch the scientific revolution of the later Renaissance. His mentor in Pisa was **Ostilio Ricci**.

Ostilio Ricci (1540–1603) was a mathematician who taught at the University of Pisa. **Galileo** took his courses and was converted by Ricci to become a scientist and mathematician. He taught Galileo Euclidian and Archimedean mathematics. Ricci believed that mathematics was not a science itself but a tool that could be used for applied science. He was mentored by **Niccolò Fontana Tartaglia**.

Niccolò Fontana Tartaglia (1499–1557) was born in Brescia. When the French defeated the Italians in that region they massacred most of its inhabitants. Niccolò was a child and was hit by a saber that sliced his jaw and palate. His mother nursed him back to health, but it made speech difficult and his nickname (the stammerer) became his last name. He never shaved, believing his beard would hide his wounds. He learned engineering and wrote books on applied mathematics. He worked out the mathematics for ballistics. He wrote a treatise on salvaging sunken ships. He translated Euclid into Italian. He solved the mathematics for cubic equations. His treatise on mathematics was a sixteenth century best seller. His most famous student was **Ostilio Ricci**.

9 My Connection to Muller's Academic Pedigree

In his academic career, Muller served as a mentor to numerous students. Most of these were graduate students, technicians, and students in his classes at Columbia, at Rice University, at the University of Texas, at the Kaiser Wilhelm Institute in Berlin, at the genetic institutes in Moscow and Leningrad when he was in the USSR, at the University of Edinburgh, at Amherst College during WWII, and at Indiana University. Two students who did not go into genetics are of interest. **Amelia Earhart** took Muller's course when she was an undergraduate at Barnard College and he was a graduate student. They became friends. She wanted Muller to take flying lessons, but he couldn't afford them. **Carl Sagan** spent a summer as an undergraduate working in Muller's laboratory. He was an undergraduate at the University of Chicago at the time. He hoped to apply his knowledge of genetics to astronomy, especially the hunt for intelligent life in the universe and the chemical stages of evolution leading to life.

Among his students at the graduate level are (in chronological order): **B. Glass, W. S. Stone, C. P. Oliver, D. Raffel, A. Prokofeyeva, C. Offerman, I. Agol, J. Kerkis, S. Levitt, R. Berg, G. Pontecorvo, C. Auerbach, A. R. Sidky, S. P. Ray-Choudhury, I. Oster, S. Abrahamson, A. Schalet, S. Iyengar, E. Carlson, W. Ostertag, S. Frye, William Trout III**, and **D. Wagoner**. Like me, all these students and the students they mentored, could attach themselves to the Muller pedigree. Colleagues who were mentored by Muller include **E. Altenburg, J. Huxley, J. Patterson, T. Painter, A. Serebrovsky, N. Dubinin, N. N. Medvedev, N. Timofeef-Ressovsky, M. Delbrück, J. D. Watson**, and **J. Crow**. At the same time, colleagues were providing skills and ideas to Muller and the exchanges were often two way. That happens also with graduate students and postdoctoral students where ideas are freely exchanged. Note that at least 35 people can be attached to Muller's academic pedigree.

While Muller was influential on my life as my primary mentor for genetics, I had a second, earlier, profound mentor when I was in Thomas Jefferson High School in Brooklyn, NY from the Fall of 1947 to the Spring of 1949. I read aloud to Morris Cohen from the classics over a period of five years, meeting him for an hour about 7 a.m. five days a week while school was in session. I attended NYU on a scholarship and majored in biology and minored in history. Mr. Cohen's influence was in relating science to society and history. It was Mr. Cohen who told me to read Schrödinger's *What is Life?* while I was still in high school. I was accepted to Indiana University in Bloomington, Indiana where I studied genetics with Nobelist **H. J. Muller**. I have had the pleasure of supervising the PhD dissertations of six students (and seeing 13 of my books published). My most noted books are *The Gene: A Critical History* (1966), *Genes, Radiation, and Society: The Life and Work of H. J. Muller* (1981), *The Unfit: A History of a Bad Idea* (2001), and *Mendel's Legacy: The Origins of Classical Genetics* (2007). My PhD was on the structure and mutability of the dumpy gene in fruit flies. My laboratory research involved comparative genetics, gene structure, and mosaicism. I applied my work on mosaicism to medical genetics using retinoblastoma as an example. I have published articles on gene structure, chemical mutagenesis, and human genetics. I taught at Queen's University in Canada (Queen's University), at UCLA, and at Stony Brook University before retiring with my wife, Nedra, in Bloomington, Indiana. Muller' concerns about the applications of genetics to society became part of my approach to writing and teaching.

10 My Academic Descendants

I have supervised the research of six PhD students and one master's student. I have also mentored dozens of undergraduates; some have gone on for PhD or MD-PhD programs. The six PhD students were all at UCLA.

1. **John Southin** (1937–2015) was born and died in Brockport, Ontario, in Canada. I met him as a student in my genetics class at Queen's University. John became my first graduate student and came with Nedra and me when we moved to Los Angeles. At UCLA John studied the mosaic distributions of induced dumpy mutations. He was a loyalist to the Queen and an admirer of left-wing rebels (Tito and Castro). He helped Americans who came to Canada to avoid fighting in the Vietnam War. He taught in Havana during the summers until he was told not to come back because he was gay. He taught the rest of his career at McGill University. He opened an androgynous bookstore in Montreal. He retired to Brockville where he died of a neuromuscular degenerative disease. John was an outstanding teacher at McGill.

2. **Ronald Sederoff** (b. 1939) was born in Montreal, Canada. He worked in my laboratory as an undergraduate and started a Medical School at Stanford but decided he preferred research. He did his dissertation on a comparison of mutagenesis in bacteriophage (with **Robert Edgar** at Caltech) and with fruit flies at UCLA. He went to Geneva to do a postdoctoral stay with

Charles Epstein. He settled in the University of North Carolina in Raleigh and worked on forestry genetics devising a technique to introduce DNA into woody tissue and culturing trees from the altered cells. This was both new and important and it led to Ron's election to the National Academy of Sciences. In 2018 Sederoff received the Wallenberg Prize for plant sciences in Stockholm from the King of Sweden. He was still active in 2017 as an emeritus professor.

3. **Harry Corwin** (1938–2017) was an able student in my genetics class at UCLA and I asked him to explore working in my laboratory. He studied chemical mutagens in Drosophila. He enjoyed his academic life at the University of Pittsburg where he became the Dean of the Honors Program. He asked me to teach in the *Semester at Sea* program (Spring 1992) for which he served as an academic dean. He retired to Colorado and died in the state of Washington of complications from diabetes.

4. **Robert J. Hendrickson** studied the cytology of the dumpy locus and induced rearrangements with X-rays that altered that gene's expression. He did a postdoctoral at Yale where he became an alcoholic. He had served in WWII and was about ten years older than me. He lost two jobs and joined Alcoholics Anonymous in Denver. He earned a living as a photographer. I used to meet him in Colorado Springs when I was active with the Lilly Endowment workshops in the Liberal Arts. He disappeared a few years after recovering from a heart attack.

5. **Dale Grace** (1939–1990) was a gymnast as an undergraduate. He studied the structure of the dumpy gene and added additional sites to its map. He went to Holland for a postdoctoral study but switched to medical school there. He dropped out after a severe case of mononucleosis. He went to Oregon and studied mosquito genetics. I last saw him at a genetics congress meeting in Toronto in 1988. He died shortly after that.

6. **John Jenkins** (b. 1941) was born in Springfield Massachusetts. He did his undergraduate work at Utah State and joined my laboratory and used ethyl methane sulfonate as a mutagen to compare chemical and spontaneous dumpy mutations. He took a position at Swarthmore College and is still active there as an emeritus professor. He has written two textbooks in genetics and human genetics.

7. **Shari Cohn** (b. 1957) was born in Plainview, New York and she was an undergraduate at Stony Brook University who worked in my laboratory for undergraduates. She did a project on color blind expression in carrier females studying one eye at a time and using Ishihara charts with diminished lighting and other means of comparing homozygous XX or hemizygous (XY) normal color vision individuals from mutant bearing carriers. She found variations in color perception in such heterozygous females exist, confirming **Mary Lyon's** hypothesis of X-inactivation for this trait. She extended this to a Master's degree with me and **David Emmerich**, a psychologist at Stony Brook University, using more sophisticated machinery, a tachistoscope, to measure the time involved in recognizing colored dots or numbers. Shari went to Edinburgh, Scotland to do her PhD on second sight exploring its folklore, history, and prevalence in families and communities. She learned Gaelic to converse with Scottish people having a tradition of second sight experiences. She still resides in Edinburgh with her family.

Among my undergraduate students at Stony Brook University who have entered academic careers are:

1. **Alfred Handler** a PhD with John Postlethwait studying fruit fly oogenesis. He took up a postdoctoral course at Caltech and works in Florida doing Dipteran research for the US Department of Agriculture in Gainesville, Florida.

2. **David B. Weiner** got his PhD in Cincinnati and did research on vaccines at the University of Pennsylvania, where he is now Vice-President for Research at the Wistar Institute. He is noted as a "father of DNA vaccines."

3. **Philip F. Giampietro** received his MD at Stony Brook University and his PhD with **Robert Desnick** at Mount Sinai Medical School in NYC. He specialized in human genetic disorders and taught in Wisconsin before moving to the Philadelphia region, where he is a professor of pediatrics at Drexel University.

4. **Daniel Ciccarone** (b. 1960) was born in New York City. He received his MD at Stony Brook University and he is now a Professor of Community and Family Medicine at the University of California, San Francisco Medical School, where he got a MPH. He has published on the prevention of HIV transmission and opioid addiction. He has testified before Congressional Committees that addictions are stigmatizing diseases and need community responses that address their social problems and not just their psychological problems. He views were shaped in his youth. When he was attending medical school and his parent went bankrupt, he slept in a tent in the woods by the Biology Buildingw and took his showers in the basement of that building.

5. **Owen Debowy** received his PhD and MD at New York University with a dissertation on the neurological control of vision perception. He is Medical Director at Sturdy Medical Hospital in Plainville, Massachusetts, with specialization in internal medicine and pediatrics.

6. **Thomas Houze** got his PhD in Gothenburg, Sweden, and worked in molecular medicine and holds a patent. He went to Great Britain and cofounded a startup company using stem cell

research. He worked for the NIH and is now with the FDA in Silver Spring, Maryland.

7. **Bruce Luke Wang** (b. 1965) got his PhD at the University of Illinois in Chicago where he worked on molecular pharmacology projects.

8. **Gary Joel Vorsanger** got his PhD and MD at Mount Sinai Medical School and has worked in drug development in the pharmaceutical industry. He is founder and President of Crossroads Scientific Medical Company, Morrisville, Pennsylvania. He specialized in internal medicine and Anesthesiology.

9. **Peter W. Thompson, MD** attended medical school and has focused on the hospital centered patient care model of treatment. It is heavily committed to maintaining a relation with patients throughout their illnesses. This physician-owned program is in some 20 states. It is called *Apogeephysicians* and mentoring is a major part of how participating physicians are trained. Thompson is Chief of Clinical Operations and mentors' young physicians in the program.

10. **Suzanne M. O'Neill** was the coordinator for my Biology-101–102 course. She took an interest in human genetics and went to Pittsburgh to study genetic counselling and eventually got a PhD there in 2001. She enjoys the practice of genetic counselling and helps other students entering that field at Northwestern University in Chicago, Illinois.

11. **Leonard Kellner** (1952–2018) did a project on twinning, taking photos of identical and same sex non-identical twins. He would cut these, and match the left face of one twin with the right face of the other twin and this showed asymmetry for non-identical twins but perfect matching for identical twins. He also got photo albums of twins and showed that at all stages of life to old age the identical twins

maintained their symmetry. After graduation Kellner took an interest in non-invasive prenatal diagnosis and founded his own company for detecting alpha-fetoprotein in maternal blood and other markers in maternal blood for a variety of genetic disorders. The alpha-fetoprotein identified a faulty development of the neural tube leading to spina bifida and anencephaly.

12. **Tracey E. Meyers** got her MD at NYU and worked for several years at Harlem Hospital working with sickle cell anemia patients and those at risk. She moved to Duke University in North Carolina and teaches Family Medicine.

From UCLA, I would add mentoring as an undergraduate, **Anthony Shermoen**, who got his PhD at Wesleyan University in Connecticut. He does research on embryonic transcription in Drosophilia at the University of California, San Francisco, California.

Many students I have mentored so that they could direct their energies to creative outlets. This includes several Stony Brook students. **Jonathan Hanke** got his PhD in mathematics and taught basic creative mathematics as a professor and now applies mathematics to industry as a vice president at Goldman Sachs. He devotes time to encouraging high school students to experience mathematics as a creative activity. **Michael Kramer** developed his talents in computers and art and became Viacom's art director. He also used his talents to restore WWII airplanes for the *USS Enterprise* museum in NYC. He now works for IBM in Texas. Michael's older brother, **Richard Kramer**, also entered a field using computers as a systems engineer in Chicago. **Howard Diamond** studied Biblical references to sexual and eugenic practices and attitudes in the Old Testament.

I learned a lot from his insights into Jewish interpretations (mostly Talmudic) of birth defects, intersex conditions, and eugenic practices. He became a Rabbi at Temple Bńai Sholom Beth David, Rockville Center, New York. **Leonard Jay Moss** started a graduate program at Albert Einstein Medical School and then shifted to a DO degree with a specialty in cardiology. He practices in New Jersey. **Michael Yeh** practices emergency medicine and toxicology. He combined journalism, epidemiology, and medicine in his formative years, and he hopes to write about medicine and society from his experiences. He is now at Emery University in Atlanta, Georgia. **Sean Li** practices anesthesiology and pain management in New Jersey and has published articles and book chapters in his field. **Steven Chaikin** became a lawyer representing impoverished clients in criminal cases. **Scott Stein** started a PhD program at IU but switched to medicine and specializes in rheumatology and immunology and practices in Victoria, Texas. **Burton Rocks** (b. 1972) got his law degree and combined it with his love for spectator sports. He coauthored autobiographies of baseball players and developed his own firm for representing them for their negotiated contracts. He teaches sports contract law as an adjunct professor at Stony Brook University. **Adam Greenberg** got his MD at SUNY Downstate and studied cell biology and genetics at Cornell Medical University. He is now at UC Davis.

I would also add a student I knew mostly through telephone conversations over more than two decades. **Mark Italiano** (1960–2017) was a gifted pianist who taught piano in Colonia, New Jersey to make a living but had an interest in what were then called hermaphroditic disorders. He earned a doctorate degree in alternative medicine and he wrote articles on intersexual disorders and how they have been interpreted by science and Society.

I would add **Anthony Delurefficio** who worked at Cold Spring Harbor Laboratory Library as an archivist for the Watson papers. I have encouraged his interests in the history of genetics. He was a Librarian for the New School in NYC and now is data managing at Sloan Kettering.

I single out the influence of **Robert Desnick** (b. 1943) on my interests in human genetics. Desnick is Dean of Genetics and Genomic Medicine at the Mount Sinai Health System, New York. He is a member of the Institute for Medicine of the National Academy of Sciences. I met him while I was a visiting professor in the history of science at the University of Minnesota and came back to spend a semester in his laboratory at the University of Minnesota and went on rounds with his pediatric fellows to acquaint myself with over 100 human genetic disorders and while there co-authored two articles with Desnick on mosaicism in retinoblastoma and how to counsel families with varied onsets of the condition in their children. Desnick and I also mentored a Stony Brook University student I recommended to him, **Steven B. Galson**, (b. 1956) who became Acting Surgeon General under the Bush Administration and Assistant Secretary of Health in the Obama administration. Galson has focused on public health issues such as obesity in children and the prevention of epidemic diseases. He is now Senior Vice President of Amgen.

Note that 26 names can be attached to my intellectual pedigree. With the 35 from Muller's intellectual pedigree, that yields 61 persons associated with the Muller pedigree.

11 Representing Inputs to My Academic Life

The circular figure that follows represents my life from the perspective of inputs. We know many of the major influences on our lives. Many of them are shared by others. Each of us is unique in the environmental sense that we live independent lives (even conjoined twins). The circle begins at noon and goes clockwise. For convenience I have listed these in more detail in the legend to this circle of inputs. I have listed the environmental factors but realize there are also genetic factors that may filter or distort our reception and interpretation of environmental influences. Those are harder to designate because our knowledge of the genetic composition and regulation of the nervous system is still inadequate compared to what we know about environmental influences on our lives. Muller used the phrase "life the lucky" to represent the diversity both genetic and environmental influences in our lives. When we are born may determine if we experience wars and catastrophes. We have no control over who our biological parents are or if we are orphans or adopted and raised by parents unrelated to us biologically.

In my own life I have been heard by thousands of students because I preferred to teach large introductory courses with 100 to 600 students in a lecture hall. I presume this is also true for the thousands of persons who have read my various books and articles. But as people leave their institutions and work elsewhere or retire or die, they are rapidly forgotten. I used to ask my students to name as many Nobel laureates in the life sciences as they could, but few could name more than a dozen. My selection of inputs is thus biased because I do not remember who taught me each fact or idea I know or when I first came to know them.

I chose Muller as my second example because next to myself, I know more about his life than I do even of my siblings. I had the privilege to write his biography and I helped classify thousands of documents he left to the Lilly Library at Indiana University. Muller's life is more adventurous than mine and he lived his adult life largely in a world at war or in economic collapse. His education reflects more of the nineteenth century than does mine. We both chose academic lives, but Muller's was much more heavily involved in research and mine was more heavily involved in teaching and writing. We both had failed first marriages and very happy second marriages, but this is probably coincidence rather than a reflection of inputs into our personalities. We both had a Jewish ancestry on our mother's side, but this too is probably coincidental. It is difficult to know what we can infer about upbringing from the inputs into our lives. But we shouldn't minimize the power of ideas, personality, and skill that lead us to emulate those who have had positive influences on our lives.

By representing inputs and outputs as circular diagrams, I bring out the complexity of how our minds are shaped. There are dozens of biographies of scientists and it will be an interesting project to do a comparative study of inputs and outputs. They may reveal some common features and the time in which shared components emerged in these lives. I believe molecular and genetic insights into creativity and mental development will eventually be revealed in detail. I also believe that studies of inputs and outputs have much to offer scholars in the history, philosophy, and sociology of science.

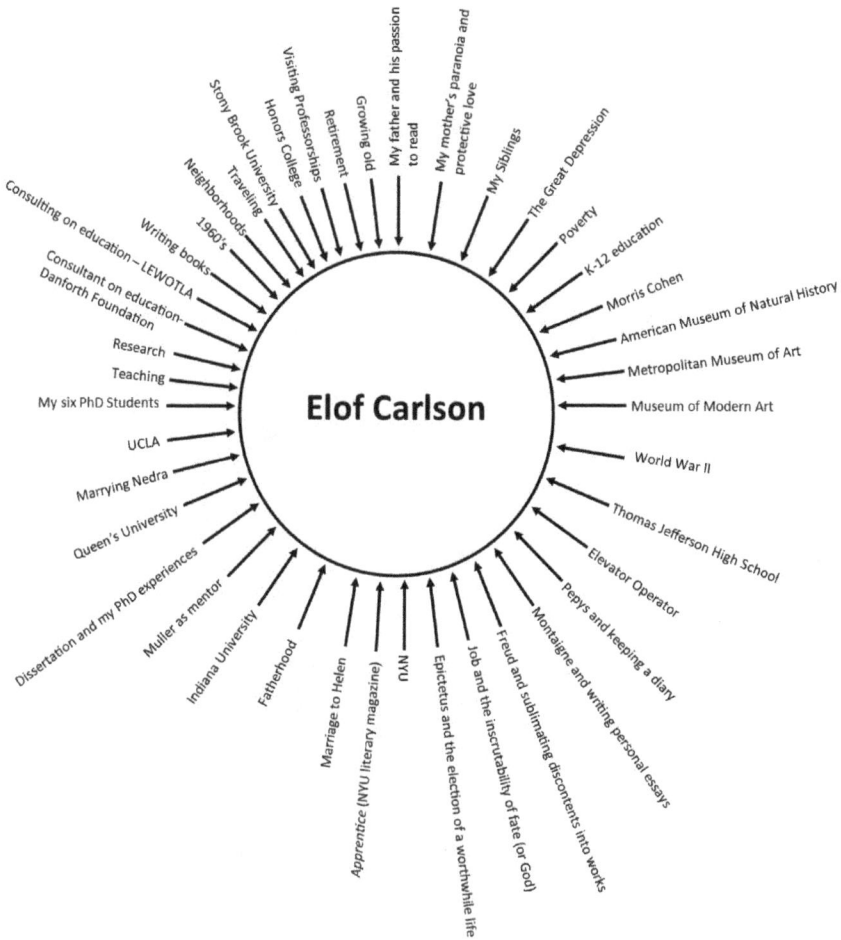

Stony Brook University
Honors College
Visiting Professorships
Retirement
Growing old
My father and his passion to read
My mother's paranoia and protective love
My Siblings
The Great Depression
Poverty
K-12 education
Morris Cohen
American Museum of Natural History
Metropolitan Museum of Art
Museum of Modern Art
World War II
Thomas Jefferson High School
Elevator Operator
Pepys and keeping a diary
Montaigne and writing personal essays
Freud and sublimating discontents into works
Job and the inscrutability of fate (or God)
Epictetus and the election of a worthwhile life
NYU
Apprentice (NYU literary magazine)
Marriage to Helen
Fatherhood
Indiana University
Muller as mentor
Dissertation and my PhD experiences
Queen's University
Marrying Nedra
UCLA
My six PhD Students
Teaching
Research
Consultant on education–Danforth Foundation
Consulting on education – LEWOTLA
Writing books
1960's
Neighborhoods
Traveling

Elof Carlson

12 List of Inputs into My Life

1. **My father and his passion to read**: My junior high school mentor, Hugh F. Browne, once responded to a get-well card I sent to him with a quotation I was told to show to my father. It was in Latin "Exegi monumentum aere perennius." (I have built a monument that is more enduring than brass). I learned it was an ode written by Horace. He was thanking my father for his contributions to my talents.

2. **My mother's paranoia and protective love**: I learned many psychotic people are not psychotic all the time. When my mother was in her sane moments, she was generous, loving, and protective and encouraged my brother and me to enjoy the arts and humanities.

3. **My siblings**: I had virtually no contact with my half-sister, Sadie, but did look forward to trips to Manhattan to visit my half-brother, Benny. I learned not all communists were terrorists with bombs, but many were good people who sought justice and equal opportunity in an era that tolerated racism, sexism, and exploitation.

4. **The Great Depression**: Growing up poor taught me that a pencil and paper or a dime store box of watercolors could be used to generate art and that NY City's policy of free museums was one of the great gifts of government to people of all social standing, even the homeless and impoverished.

5. **Poverty**: I share with most of my classmates the poverty of the Depression. Many were less fortunate and were from homes where fathers were out of work and they depended on "relief" as it was called then and "welfare" as it is called today.

6. **K-12 education**: I bless the Board of Education of New York City for its wonderful teachers who cared for their students and went out of their way to help students learn. One teacher even gave me the class Christmas tree to take home.

7. **Morris Cohen**: I owe so much to Mr. Cohen. What a gift he gave me to read aloud the major works of Western Civilization.

8. **American Museum of Natural History**: My favorite museum. I devoured the legends of the display cases as I went dozens of times each year on weekends or during the summers to enjoy its displays of the artifacts of the universe.

9. **Metropolitan Museum of Art**: I was captivated by the art of all ages. Looking at bas-reliefs carved in the wooden backing of an Egyptian chair from a tomb was stunning in its detail. I realized these people living some 4000 years ago were as talented as we are today, and I felt a communion with them just enjoying their skills.

10. **Museum of Modern Art**: I was amazed at the limitless ways human imagination could transform the familiar. Reality was constantly transformed, especially in human portraiture.

11. **World War 2**: I was grateful the allies defeated the Axis powers as we called them. But I felt sad that everything Japanese was rejected. Before the war I looked lovingly at the Bonsai miniature trees in the *Encyclopedia Britannica* and appreciated the sensitivity of the Japanese artists and architects with a set of photo postcards of Kyoto I got from my mother when I admired them in an antique shop.

12. **Thomas Jefferson High School**: I loved being in high school with wonderful teachers in English (Mr. Felix Sper), French (Mr. Max Cantor), earth science (Mr. Benjamin Schupack), mathematics (Ms. Edna Kramer), and art (Mr. Bernard Green). Mr. Cohen's outreach to my education was a gift that would be hard to duplicate even for a millionaire's child.

13. **Elevator operators**: I learned what hard work is for most people and I appreciated the banter with my fellow elevator operators during my summer jobs.

14. **Pepys and keeping a diary**: Reading Pepys told me how to learn history from the eyewitnesses of those experiencing it. He also taught me the discipline of doing so every day. I wish I had had the time to do since 1949 to the present, but at least 80 percent of my life is represented in over 100 volumes of a diary.

15. **Montaigne and writing personal essays**: His essays talked to me about the pain he suffered from kidney stones (something I learned to experience with my own calcium oxalate stones). I felt like he was talking to me. Personal essays are a way to experience a life after death not based on faith but on the reality of pen and ink.

16. **Freud and sublimating discontents into works of civilization**: Despite the Freud bashing since the 1960s, I learned from reading his *Civilization and its Discontents* how to convert ("sublimate") frustration, failure, or melancholy into works of art, science and scholarship. I wish every student would read that book in high school or in the frosh year at college.

17. **Job and the inscrutability of fate (or God)**: I never believed in a God because the trips to museums revealed so many dead religions and changing views of God and gods.

But I admired the author of the Book of Job for creating a character who had the courage to challenge God, punishing Job for a lousy bet with Satan and bullying Job into groveling submission by telling him, "Who the hell are you to question my judgment?" If I had been raised religious, that alone would turn me away from it.

18. **Epictetus and the election of a worthwhile life**: I learned that if a slave can become an author of a work that has lasted some two thousand years, that most of humanity has the capacity to surmount the setbacks or circumstances of their birth and be creative.

19. **NYU**: I experienced wonderful teachers — **Charles Davis** in English, **Wallace K. Ferguson** in history, **Thomas J. King** in biology.

20. *Apprentice* (**NYU literary magazine**): I learned how to proofread articles and put a magazine together. I had great friendships and spirited rivalries with poets and writers who enjoyed the rivalry and encouragement we provided each other.

21. **Marriage to Helen**: I learned from Helen her commitment to poetry and I failed to be her muse. But I appreciated being demoted to a friend rather than being written off as a failure.

22. **Fatherhood**: I had the joy of Claudia with Helen and the joys of Christina, Erica, John, and Anders from Nedra. I believe my personality, my joy in their presence and the love and tolerance I showered on them, was appreciated.

23. **Indiana University**: It was like experiencing a second life. I was on my own. Every graduate course was an input of new knowledge by committed teachers and an atmosphere that sustained intellectual growth.

24. **Muller as mentor**: working for a Nobel laureate inspired me to emulate his habits of learning and scholarship. I admired his belief that science is not only a joy to understand and participate in but also had its moral obligations to society.

25. **Dissertation and my PhD experiences**: Spending two years of research to interpret the structure of a gene by designing genetic stocks, inducing mutations of that gene with X-rays, and reading every paper ever published on that gene (dumpy or truncate) in *Drosophila melanogaster* was like putting a key in a lock and watching a door open to the universe.

26. **Queen's University**: teaching in Canada. I learned what it feels like to work in another country and experience its students and culture. I learned to write review articles and explore my study of the dumpy locus to cognate fields.

27. **Marrying Nedra**: So far, we have experienced 60 years of marriage and Nedra is my best friend, the love of my life, and we admire each other.

28. **UCLA**: I had six students complete their PhDs and learned how to teach undergraduates. I witnessed the 60s when campus rebellion was both a threat and an opportunity for positive change.

29. **My six Ph D students: John Southin, Ron Sederoff, Harry Corwin, Robert Hendrickson, Dale Grace, and John Jenkins** all became part of my extended family as I shared my mentoring with them.

30. **Teaching**: I learned to teach by applying the skills of science to effective delivery, organization, and content of each lecture.

31. **Research**: The dumpy locus gathered complexity and I studied chemical agents to induce mutations in that region. I learned how much time it takes to apply for grants to help support graduate research.

32. **Consultant on education — Danforth Foundation**: I helped select winners of Danforth Fellowships a much-coveted award for those studying for their PhDs. I also participated in other educational projects which gave me a view of academic diversity in the US.

33. **Consulting on education — LEWOTLA**: The workshops in the liberal arts provided by the Lilly Endowment exposed me to new educational trends and the stimulation of participating in faculty led seminars for two weeks a year at Colorado Springs.

34. **Writing books**: so far 13 have been published and I have about 15 more in various stages of development. Each book taught me new resources to explore and stimulated ideas for additional books.

35. **1960s**: Most of it was spent at UCLA and I got to know some of the students involved in the student strikes and the frustrations that young people feel for a failed generation that ignores their cries for change based on equality of opportunity rather than ideologies of the left or right for power.

36. **Neighborhoods**: Slums (410 Sheffield Avenue in Brooklyn), my own home with a mortgage (Sunnyside Avenue in Mar Vista, California), (a solid middle-class split-level home in our neighborhood for 19 Mud Road in Setauket, NY, formerly occupied by the Dean of Stony Brook University). Our present down-sized one level home in Bloomington, Indiana, facing the challenges of advanced age. Travelling: Flights to many campuses to give lectures, to travel by ship twice around the world, to experience the diversity of America and the world.

37. **Stony Brook University**: A wonderful pioneering effort in a relatively new school where innovation was appreciated in teaching, research, and campus expansion.

38. **Honors College**: the opportunity to mentor very bright students and to provide opportunities for them to explore and develop their talents.

39. **Visiting Professorships**: San Diego, Salt Lake City Utah for the University of Utah, the University of Minnesota, Tugaloo College. I felt like Walt Whitman celebrating American diversity.

40. **Retirement**: a marvelous opportunity to be a full-time writer and I am still writing in my late 80s.

41. **Growing old**: make the most of it. In the almost 20 years since I retired, I have enjoyed renewing myself with writing projects. As for the physical effects of decrepitude, I don't wish to be a spoil sport. You'll learn to cope with each decade after sixty.

13 Assessing Inputs into My Life

I was born during the Great Depression in 1931. I had a brother born in 1929 who had a congenital heart defect that was non-operable in those days. He was given six months to live and lived to be 45. My father was an elevator operator who left Sweden for the Merchant Marine and then settled in NY as an elevator operator. My father had a passion for reading books and our apartment was filled with books. My mother was born in Bound Brook New Jersey of immigrant Ukrainian parents who were orthodox Jews. My father was a lapsed Lutheran who was an atheist. My mother was disowned by her family. More troubling, she was schizophrenic and had been institutionalized after the failure of her first marriage which was arranged by her father. My brother and I were never given any religious affiliation and grew up as atheists. My mother's first marriage produced my half-sister and half-brother. I was born when my half siblings were in their late teens and on their own. Because of poverty and my mother's paranoia, we moved a lot and I went to eight different public schools. It was not until junior high school that I bonded with Hugh F. Browne, my science teacher. He liked my work and personality so much that he asked my parents if he could adopt me, so I would have a better chance of attending college. They refused but thanked him for his concern. In high school I bonded with M. G. Cohen and read aloud to him from classical literature from my senior year in high school to my graduation from NYU. Mr. Cohen gave me the education that millionaire's children got from private tutors in the 17th and 18th centuries. I read aloud to him because he was going blind from Leber's optic atrophy. We began with plays of Aeschylus, Sophocles and Euripides and ended with Freud's *Civilization and its Discontents*. It made me aware of the continuity of knowledge from antiquity to the present.

We were a nuclear family with no visiting relatives and no friends or acquaintances of our parents visiting our apartment. At NYU I particularly liked Charles Davis, my first Afro-American teacher and my English instructor, who encouraged me to write. I also liked Wallace Ferguson and his approach to history, so I minored in history while majoring in biology. I wrote to Muller and asked if I could be his student. I was a teaching assistant and then National Science Foundation fellow getting my PhD. Muller had enormous influence on shaping my habits as a scientist and seeing genetics in historical, social, and contemporary context. Muller's laboratory also was a place of stimulating discussions with my fellow graduate students especially Seymour Abrahamson, Abe Schalet, Sarah Frye and Irwin Oster. I benefitted too from IU's faculty whose courses I took in zoology and botany. I also learned to teach by teaching Sonneborn's Introductory Genetics Course while he was on sabbatical leave. I taught 300 students as my first lecture course while assessing my successes and failures in communicating. While in Bloomington I married Helen Zuckerman and we had a daughter, Claudia, who was born there. The marriage failed. I later met Nedra Miller and invited her to take my summer course in genetics. We corresponded while she was in Chicago and I was teaching at Queen's University in Canada. We got engaged and married in Nedra's hometown of Rochester, Indiana. We have enjoyed 60 years of a happy marriage.

My teaching shifted to teaching non-science majors during my experiences in the 1960s when students revolted against the Vietnam War and shallow values of consumerism. I introduced a course for non-majors at UCLA and then revised it when I joined the Stony Brook faculty.

At the same time as I was giving lectures in large auditoriums, I was realizing the importance of mentoring students individually both for their research and their struggles finding careers and self-confidence. I was very insecure from my teens into my thirties because of the circumstances of my birth and my parents' failures to find satisfaction in their lives.

I keep a diary and have over 100 volumes since I started in high school as a senior after reading from Pepys's diaries in my Honors English class with Dr. Felix Sper. I love writing personal essays which fascinated me when Mr. Cohen had me read several of Montaigne's essays.

14 Assessing Outputs in My Life

If a circle of inputs represents how I shaped my life. A circle of outputs represents how I shaped the lives of those around me. Each of us has this dual set of inputs and outputs that make us who we are or were. As an old man in his late 80s after the writing of this book, I have time to reflect over four generations of life. The outputs begin with birth and end with death (or even later) but by representing my life as an output circle I can assess how I evolved and what my own contributions were.

As a child I had no concept of what it is to be a geneticist, a historian of science, a scholar or a writer of many books. Those ideas began to emerge as I entered my teens. In junior high school I knew science was my learning experience, but art was a great aesthetic experience. I resolved that in high school by a careful study of portraits at the Metropolitan Museum of Art where I could see what it took to shift me from a "Sunday painter" into a portrait artist. I did not want to invest that time into painting. The pull of science and my visits to the American Museum of Natural History were more frequent as were my readings of the history of science.

I think on my life as a series of metamorphoses like an insect going through larval and pupal stages. Each university I taught in had different things to challenge me and led me to try new things to learn or write about or gain new insights into effective teaching.

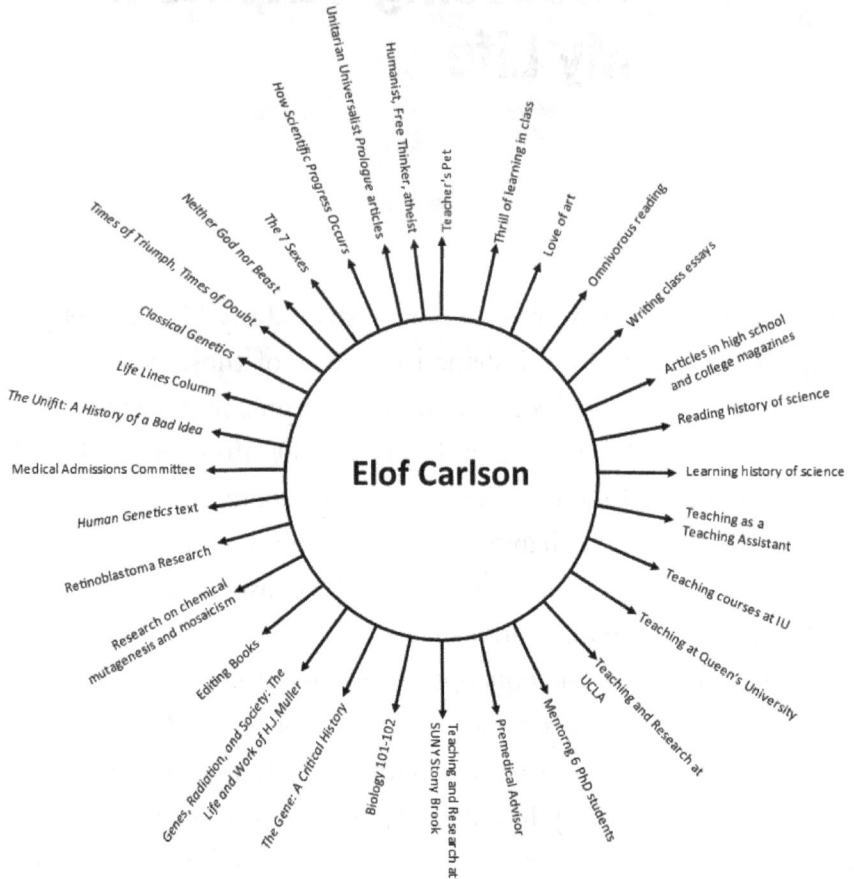

Elof Carlson

Unitarian Universalist prologue articles
Humanist, Free Thinker, atheist
Teacher's Pet
How Scientific Progress Occurs
Neither God nor Beast
The 7 Sexes
Times of Triumph, Times of Doubt
Classical Genetics
Life Lines Column
The Unfit: A History of a Bad Idea
Medical Admissions Committee
Human Genetics text
Retinoblastoma Research
Research on chemical mutagenesis and mosaicism
Editing Books
Genes, Radiation, and Society: The Life and Work of H.J. Muller
The Gene: A Critical History
Biology 101-102
Teaching and Research at SUNY Stony Brook
Premedical Advisor
Mentoring 6 PhD students
Teaching and Research at UCLA
Teaching at Queen's University
Teaching courses at IU
Teaching as a Teaching Assistant
Learning history of science
Reading history of science
Articles in high school and college magazines
Writing class essays
Omnivorous reading
Love of art
Thrill of learning in class

List of Outputs from My Life

1. **Teacher's pet**: When called on, I usually had the right answer or a commentary with it that I got from reading the *Encyclopedia Britannica* at home.
2. **Thrill of learning in class**: I loved school. It was like a sanctuary and a never-ending source of new knowledge.
3. **Love of art**: I did a lot of drawing and explored doing portraits and sketching scenes, but the demands of life and my career made me abandon that effort.

4. **Love of science**: Learning how the universe works and contributing, even one small bit of knowledge about it, was the motivating force of my life.

5. **Omnivorous reading**: It enriched my teaching, my outlook on life, my appreciation of colleagues.

6. **Writing class essays**: Teachers in high school used to read my essays aloud to the class.

7. **Articles in high school and college magazines**: I enjoyed seeing my essays in print and the compliments from fellow students and teachers inspired me to do more.

8. **Reading history of science**: Everything has a history and I learned to personalize that knowledge in my lectures with anecdotes I had heard or read of great scientists and their discoveries. I realized that writing books on the history of science was another outlet to reach students and colleagues.

9. **Learning history of science**: My research habits applied to the history of science and I combined interviews, primary sources, and a good retentive memory helped jar the memories of those I interviewed for my books on the gene and for Muller's biography.

10. **Teaching as a Teaching Assistant**: It was my first opportunity to be a teacher. We were told not to lecture but I couldn't resist smuggling in tidbits of history or anecdotes when going around the room, table to table, and discussing the objects students were dissecting or describing in their notebooks.

11. **Teaching courses at IU**: My first lecture class in 1957 was the major's genetics course. I had 300 students. I learned from faculty I consulted and from my own research on what worked and what failed. By the end of the course I knew I had the makings of a much-appreciated professor.

12. **Teaching at Queen's University**: It was a delight to learn each topic and present it to my students. Only by teaching did I appreciate how to present it and what would make it come alive to the students. My Canadian students even came in at night to help me mimeograph and staple the handouts. They also got me drunk on Canadian beer because I did not know it had a higher alcohol content than American beer.

- **Teaching and research at UCLA**: I learned at UCLA to communicate to different levels of students: graduate seminar courses, senior level lecture courses, interdisciplinary courses, team taught courses for introductory biology for majors, and a non-majors course for several hundred students.

13. **Mentoring six PhD students**: Each did worthy dissertations, but each responded differently in their academic careers. **John Southin** returned to Canada and was an outstanding teacher. **Ron Sederoff** sublimated his energies into research on forestry genetics and was elected to the National Academy of Sciences and won a Wallenberg Prize in 1957. **Harry Corwin** became a Dean at the University of Pittsburgh. He died of diabetes. **Robert Hendrickson** lost his career (and life) to alcoholism, **Dale Grace** was enjoying research on mosquito genetics in Oregon and died prematurely. **John Jenkins** became a successful teacher at Swarthmore and has written two texts.

14. **Premedical advisor**: I served in this role for two years at UCLA and served more than ten years as a member of the medical admissions committee at Stony Brook University. I learned how diverse premedical students are and felt reassured that merit, service, motivation, and talent were all assessed in both roles.

15. **Teaching and research at Stony Brook University**: I shifted to a laboratory for undergraduate research so I could

stress undergraduate teaching. This gave me the opportunity to design a non-majors course in biology and to have time to do research in writing books in the history of genetics.

16. **Biology 101–102**: I taught this course from 1968 to 2000 at Stony Brook University. It usually had 500 to 600 students per semester. It was a highly successful course and led to my receiving a $10,000 award from the Danforth Foundation for their Harbison Award for Gifted Teaching.

17. ***The Gene: A Critical History***: This was my first published book. It established my reputation as a scholar in the history of genetics. I explored the clash of views on the gene that led to new fields to reconcile opposing views.

18. ***Genes. Radiation, and Society: The Life and Work of H. J. Muller***: I used the 20,000 documents at the IU Lilly Library to write Muller's biography and used the money from my teaching award to fund travel to Europe and around the US to do interviews for the Muller biography.

19. **Editing books**: I edited a book on the history of biology for the Book Find Club. I edited two volumes of Muller's essays. I also edited a collection of papers on the history of the gene concept.

20. **Research on chemical mutagenesis and mosaicism**: I showed that most X-ray induced dumpy mutations were "complete" mutations (all eligible tissue for expressing it did so). But the overwhelming number of induced dumpy mutations by quinacrine mustard were mosaics (parts of the embryo did not carry the induced mutation). This suggested the mutated sperm had one, not both, nitrogenous bases in a base pair altered by the chemical agent.

21. **Retinoblastoma research**: My work on mosaicism with chemical mutagenesis suggested spontaneous mutations

might also produce some mosaic mutations. I teamed up a sabbatical leave with a sojourn with Robert Desnick at the University of Minnesota and we published two articles on this analysis in which Desnick obtained all the files on patients at the University Hospital on retinoblastoma for me to analyze.

22. **Human Genetics text**: I used my *Human Genetics* text for my Biology 101–102 course because the editor at D. C. Heath felt he could sell a book to non-majors with that title, but it would not be accepted as a book on biology for non-science majors with the title of *Biology: A Humanities Approach*. It was too radical for that era's expectations.

23. **Medical Admissions Committee**: Those were enjoyable years with about five hours of interviews a week and two hours of discussion and voting each week.

24. **The Unfit: A History of a Bad Idea**: I began this book at Indiana University while on sabbatical leave and continued the research at Cold Spring Harbor Laboratory archives where most of the primary sources were housed. Jim Watson read the first draft of the completed book and recommended to CSHL Press that they publish it. It made Watson my patron saint!

25. **Life Lines column**: In 1997 I sent some personal essays that I called *Life Lines* to Leah Dunaief, the publisher of the local paper, *the Village Times*. About three months later she sent me a note that she read them, liked them, and would publish them every other week. I have written about 600 of these essays (about 500 words each) and they are my effort to teach the relation of science to values, politics, culture, history, and significance in our lives. I continue to do these but since 2017 they are twice a month. They appear in six of her newspapers from

Wading River to Cold Spring Harbor on the North Shore of Long Island.

26. **Mendel's Legacy: A History of Classical Genetics**: This is a history of classical genetics which I enjoyed preparing, mostly from the Davenport and Demerec collection of reprints at Cold Spring Harbor Laboratory.

27. **Times of Triumph, Times of Doubt: Science and the Battle for Public Trust**: In this book I use examples of issues involving science and how users of applied science respond to critics of its usage. I show the science needed to make decisions and how they are often related to political debate.

28. **Neither God nor Beast: How Science Changes Who We Think We Are**: I argue that science greatly benefits humanity with the germ theory, year-round wholesome nutrition, environmental protection laws, the use of science to expose pseudoscience, and the shift in our thinking from wishful thinking to reasoned policies. These are changing our world view and our concept of human nature.

29. **The 7 Sexes: The Biology of Sexuality**: I show how each component of our sexuality was identified over the centuries so we can now talk of chromosomal, genetic, hormonal, internal genital, external genital, gonadal, and psychological sex.

30. **Mutation: From Darwin to Genomics**: Mutation was identified as variation in contrast to heredity which was identified as a species characteristic. That changed in the 19th century as evolution, plant breeding, studies in hybridization, and studies of pathological variations accumulated leading to mutation as a change in the individual gene and a host of chromosomal and epigenetic components associated with short-term and long-term transmission.

31. ***How Scientific Progress Occurs: Incrementalism and the Life Sciences***: I argue that it is not paradigm shifts that prevail in the life sciences, but new tools, more data, and experimental approaches that lead incrementally to new fields, fusion of fields, replacements of fields, and the emergence of what are called scientific revolutions.

32. **Unitarian Universalist *Prologue* articles**: I write a 250-word column every other week in the Unitarian-Universalist Church of Bloomington, Indiana newsletter (*The UUCB Prologue*) on the history of this movement from antiquity (going back to Pharaoh Akhenaton) to the present. Much of it deals with the evolving concept and role of religion in human life in both its positive and harmful ways.

33. **Humanist, Free Thinker, atheist**: I am an atheist, Humanist, and Unitarian in my religious outlook. I reject the supernatural in any of its forms (soul, spirit, God, oversoul, miracles, witchcraft, or ghosts). I favor a humanism that respects reason, appreciates the contributions of science to our world view, and encourages humans to act in ways that respect diversity, help others less fortunate, reject bigotry, shun violence, and provide hope that each generation tries to better life for its coming generation.

34. **Constructing Intellectual Pedigrees** (this book): I was happily surprised that I could trace my mentoring history back to Newton and Galileo. By doing many of these I have learned how knowledge moves from country to country and how the university has changed since the Renaissance. There are few autodidacts and few who supported their studies with their own wealth or income. I believe this book will stimulate an interest in the sociology of science as other fields try this mentoring approach.

15 Inputs into Muller's Life

The circular representation of Muller's inputs reveals his father's influence in introducing evolutionary science to him at an early age and a liberal or socialist outlook for shaping a better world. The early death of his father shifted Muller's sources of knowledge and world view to his teachers and extended family. The Mullers and Kroebers were middle class for several generations and liberal in outlook since the 1848 revolution that sent them to the United States from Germany. Muller's inputs at first were for chemistry and physics as "hard" sciences. Except for evolution, biology was largely descriptive with classification and anatomy prevailing. A scholarship to Columbia University placed Muller among the leading biologists of the turn of the century. E. B. Wilson and T. H. Morgan both provided mentoring, Wilson for microscopy and cell biology and Morgan for experimental embryology. 1906 was a crucial year when Wilson urged Muller to read R. H. Lock's book on Mendelian heredity. Muller read it as a bellhop summer employee in a hotel in Ocean Grove, NJ. Columbia was also the place where the fly lab emerged, Muller being a contributing member especially through discussions with A. H. Sturtevant, C. B. Bridges, and E. Altenburg, Muller's lifelong friend from high school.

Muller attributes his socialism to reading *The Masses* and visits with Bridges to the Greenwich Village circle of socialists cultivated by Bridges. At Rice he was influenced by Huxley with whom he had long walks and discussions on evolution. His colleagues at the University of Texas extended his research findings in radiation genetics. He also married a mathematician, Jessie Jacobs and had a son, David. Muller also suffered from depression. He said he had a breakdown about 1912 or 1913 from overwork (he supported his widowed mother). In 1932 he attempted suicide as his marriage failed, his work was attacked, and the FBI was following his participation in what were deemed subversive Communist activities, such as editing an underground newspaper for socialist students. Muller was fortunate to have won a Guggenheim fellowship and went to Berlin to work in the Kaiser Wilhelm Institute with N. V. Timoféev-Ressovsky. Muller explored the limits of using X-rays to measure gene size and structure. He also met biophysicists and physicists (like Max Delbrück) who hoped to explore physics and genetics. Hitler's rise to power forced Muller to leave and join Russian geneticists in Leningrad through the invitation of I. V. Vavilov. Things went well for three years and then a growing anti-genetic, Lamarckian movement under Lysenko emerged in the USSR. Once again Muller was forced to leave after two of his students were arrested and were shot as Trotskyites. Muller volunteered to fight in the Spanish Civil War. He then found refuge in Edinburgh, Scotland. His work there was highly productive, and he had several talented students, especially Guido Pontecorvo and Charlotte Auerbach. Muller's views on eugenics were mixed. He looked upon the US movement as bigoted, sexist, and tilted to the privileged. Muller favored positive eugenics getting his initial inspiration as a youth from reading Francis Galton's work. Also, at Edinburgh Muller married Dorothea Kantorowicz and after returning to the US, they had a daughter, Helen.

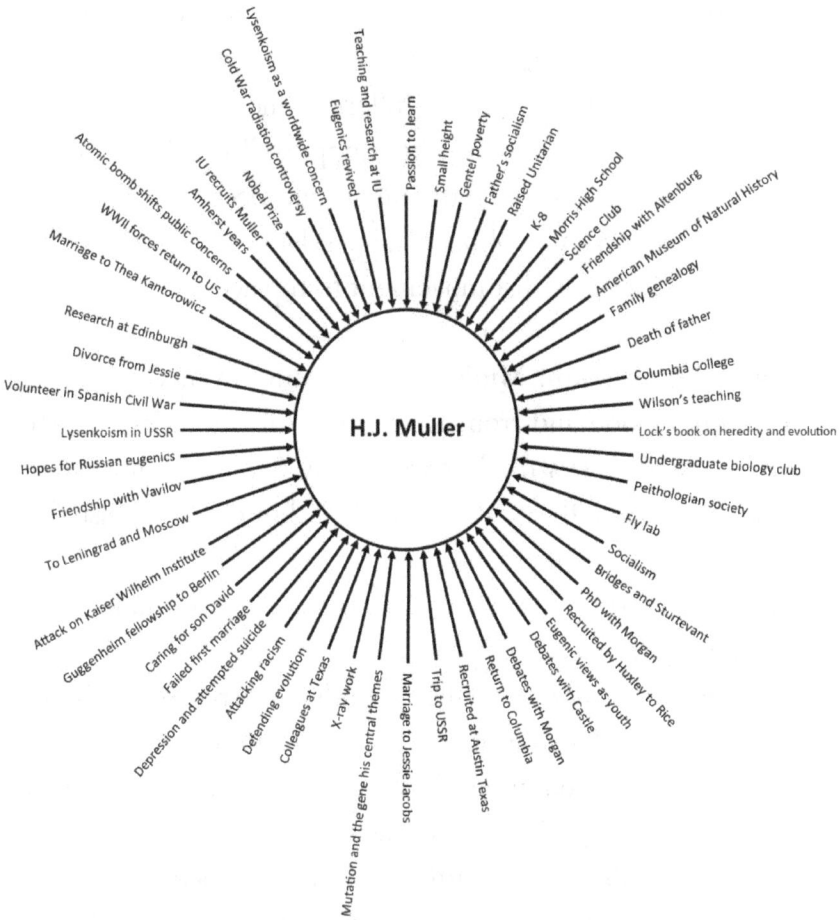

The diagram is a radial "sunburst" chart with "H.J. Muller" at the center and the following inputs radiating outward:

- Atomic bomb shifts public concerns
- WWII forces return to US
- Marriage to Thea Kantorowicz
- Research at Edinburgh
- Divorce from Jessie
- Volunteer in Spanish Civil War
- Lysenkoism in USSR
- Hopes for Russian eugenics
- Friendship with Vavilov
- To Leningrad and Moscow
- Attack on Kaiser Wilhelm Institute
- Guggenheim fellowship to Berlin
- Caring for son David
- Failed first marriage
- Depression and attempted suicide
- Attacking racism
- Defending evolution
- Colleagues at Texas
- X-ray work
- Mutation and the gene his central themes
- Marriage to Jessie Jacobs
- Trip to USSR
- Recruited at Austin Texas
- Return to Columbia
- Debates with Morgan
- Eugenic views as youth
- Recruited by Huxley to Rice
- PhD with Morgan
- Bridges and Sturtevant
- Socialism
- Fly lab
- Peithologian society
- Undergraduate biology club
- Lock's book on heredity and evolution
- Wilson's teaching
- Columbia College
- Death of father
- Family genealogy
- American Museum of Natural History
- Friendship with Altenburg
- Science Club
- Morris High School
- K-8
- Raised Unitarian
- Father's socialism
- Genteel poverty
- Small height
- Passion to learn
- Teaching and research at IU
- Eugenics revived
- Nobel Prize
- IU recruits Muller
- Amherst years
- Cold War radiation controversy
- Lysenkoism as a worldwide concern
- Atomic bomb shifts public concerns
- Marriage to Thea Kantorowicz

List of Muller's Inputs

1. **Passion to learn**: Muller loved going to the American Museum of Natural History and his father encouraged his interest in evolution.

2. **Small height**: Muller was five foot two inches tall and often had to wear boys clothing to find a decent fit.

3. **Genteel poverty**: Muller's father died when he was about 10 years old and he, his mother, and his older sister depended

on Muller's uncle to provide support until Muller and his sister were old enough to work.

4. **Father's socialism**: The Mullers came to the US from Coblenz, Germany, where they supported the 1848 socialist revolutions. They had to emigrate to the US to avoid persecution. The experience of his Muller relatives instilled Muller's sympathy for a socialist movement to replace an indifferent if not exploitive capitalism.

5. **Raised Unitarian**: Muller's family came from Germany on his father's side and from England on his mother's side (the Lyons family). His mother was partly Sephardic Jewish and Anglican. When the Mullers married, they chose to raise their two children as Unitarians.

6. **K-8**: Young Muller excelled in school and lived a middle-class life during those formative years.

7. **Morris High School**: Muller attended Morris High School in the Bronx, walking across the Bridge from Manhattan to the Bronx. He excelled in his classes and thought he would enter some field of experimental science.

8. **Science Club**: Muller participated in a science club in high school and in college. It led to his friendship with other students who pursued careers in science, especially Edgar Altenburg.

9. **Friendship with Altenburg**: Altenburg worshipped Muller who shared all his ideas and goals with him. They corresponded when they had separate careers and about 400 handwritten letters exist at the Lilly Library at IU.

10. **American Museum of Natural History**: Muller cited the evolution or horses' feet as a major inspiration for his eugenic views, believing if nature could do this with horses, why couldn't humans take charge of their own evolution using reason, not natural selection, as the basis for mating choices?

11. **Family genealogy**. None of the Muller family of his youth considered themselves Jewish. But Muller acknowledged it as part of his genealogy and attributed that maternal heritage to his love for intellectual activities and science.

12. **Death of father**: Muller's father had hoped for a career in international law but had to settle for co-ownership of an Art Metal company in lower Manhattan that made lamps, sculptings, frames, and other household artistic works. It made Muller seek jobs on and off campus part time while he was an undergraduate. He supported his mother until her death.

13. **Columbia College**: Muller won a scholarship to Columbia University and was fortunate he was a winner (there were only two scholarships available at that time, one created by Pulitzer and one created by the Cooper-Hewitt Foundation). Muller won the Cooper-Hewitt.

14. **Wilson's teaching**: Muller most appreciated the teaching of Edmund B. Wilson as an undergraduate. Wilson developed the chromosome theory of heredity and was encyclopedic in his knowledge and meticulous in the preparation of his lectures.

15. **Lock's book on heredity and evolution**: Wilson recommended Muller to read R. H. Lock's 1906 book on *Heredity, Variation and Evolution* (the first text in genetics mentioning Mendel's work).

16. **Undergraduate biology club**: Muller and Altenburg joined the Biology Club and there met Calvin Bridges and Alfred Sturtevant. They took turns discussing scientific papers they read and work they were doing.

17. **Peithologian Society**: Muller also had a broad interest in society and how citizens could bring about valuable changes in society. The Peithologian Society encouraged students to dis-

cuss controversial topics. Muller presented his first ideas on eugenics to that group and his belief that it could only occur in a socialist society.

18. **Fly lab**: Muller finished his BA in 1910 before Morgan set up his fruit fly laboratory. Morgan had been an embryologist and switched to heredity after 1905. His first mutant variations with fruit flies were not found until 1909 and his work on white eyes was not published until 1910–1911. It was not until 1912 that Muller asked Morgan if he could join his laboratory.

19. **Socialism**: Muller said he became a socialist through his father and reading *The Masses*, a socialist magazine. He never abandoned the superiority of socialism over unregulated capitalism, but he later abandoned its extreme form, Communism.

20. **Bridges and Sturtevant**: Muller, Altenburg, Bridges, and Sturtevant discussed all of Morgan's work with fruit flies and they shared each other's interpretations of each new experiment.

21. **PhD with Morgan**: Muller's PhD was on crossing over (found by Morgan and mapped by Sturtevant). Muller introduced coincidence and interference as more accurate ways to map genes on a chromosome.

22. **Recruited by Huxley to Rice**: In 1915, as Muller was finishing his dissertation, Julian Huxley visited Morgan and asked if he had a student for the new Rice Institute where he would chair the Department of Biology. Morgan recommended Muller. He and Huxley formed a lifelong friendship. At Rice they took walks together and discussed their theories of evolution.

23. **Eugenic views as youth**: Muller's ideas at Columbia were largely based on Galton's ideas of positive eugenics or the encouragement of mating between individuals with high talent, good health, longevity, and attractive personalities. At Rice he felt eugenics would work best if it rejected the views of

Davenport and Laughlin which stressed negative eugenics (the reduction of alleged bad genes by sterilizing the unfit, banning them from marriage, or institutionalizing the psychotic, the recidivist criminal, and the mentally retarded).

24. **Debates with Castle**: Muller believed the gene was intact and not modified when it passed through a heterozygous stage. Castle believed in "allelic contamination" and they disputed each other in ad hominem attacks in published papers. Eventually Castle conceded that Muller's view of modifier genes for the variations he saw was correct.

25. **Debates with Morgan**: Muller found Morgan disorganized as a teacher and often contradictory or confused in his articles. He gave a talk on "Erroneous assumptions regarding genes" that criticized his mentor's work. Morgan was less irked over this attack from his own student than in rejecting his advice not to publish an attack on Castle's work. Muller did anyway.

26. **Return to Columbia**: While Morgan was on sabbatical leave in 1920 Muller taught his courses, hoping to get a position there. He formulated his ideas on mutation as a change in the individual gene there and analyzed the time of origin of the various alleles of the white-eyed gene.

27. **Recruited at Austin Texas**: After Morgan returned, Muller got a job at the University of Texas. He began studying ways to study the gene's size, frequency of spontaneous mutation, and functional properties. He also developed many stock designs to isolate specific genetic events and to maintain stocks that would otherwise get lost.

28. **Trip to USSR**: In 1922 he and Altenburg took a trip to Europe and the USSR and brought fruit flies to the Russian laboratories they visited. He wrote an article about advances in genetics in the USSR.

29. **Marriage to Jessie Jacobs**: Muller married a faculty member, Jessie Jacobs, a Kansas born and raised mathematician.

30. **Mutation and the gene** his central themes: Muller argued that polyploidy, aneuploidy, and chromosomal rearrangements of all kinds (inversions, translations) do not play as decisive a role in Darwinian evolution as do the mutations in individual genes that regulate, alter, or bring about novelty in the gene.

31. **X-ray work**: In 1926 Muller read all the literature on radiation and its biological effects and designed a stock to detect X-linked recessive lethal mutations. He called his stock ClB. Using this he applied X-rays (estimated at about 5000 roentgens) and obtained numerous X-linked genes and X-linked visible mutations all of which he mapped. He published his results in 1927 and it created a sensation.

32. **Colleagues at Texas**: Muller worked with J. T. Patterson and T. S. Painter on several fruit fly experiments and had a flourishing program for graduate students.

33. **Defending evolution**: Muller gave lectures defending evolution from the attacks by fundamentalists. It was not easy to defend Scopes and the right to teach Darwinian evolution in public schools in a University facing the State Capital and its mostly fundamentalist representatives.

34. **Attacking racism**: Muller defended blacks accused of raping white women and objected to the conditions of blacks in the south. He published his views in a student newspaper, *The Spark*, that was identified by the FBI as run by Communists.

35. **Attacking negative eugenics**: Muller felt the Eugenics Record Office at Cold Spring Harbor was racist, sexist, and elitist in a distorted way. He believed American eugenics publications downplayed the negative effects of poverty, low wages, slum-like dwellings, malnutrition, and other factors

that prevented the neglected classes from having opportunities taken by granted by the middle class and wealthy supporters of eugenics.

36. **Caring for son David**: Jessie had postpartum depression, combined with disappointment that the Mathematics department fired her because she had a child to raise.

37. **Depression and attempted suicide**: Muller felt his marriage collapsing, his colleagues competing with him for students and financial support, and the FBI closing in on him for his role in editing *The Spark*. He walked off into the woods to overdose on sleeping pills and fell asleep. When he did not show up at home or work, a posse of faculty and students found him walking dazed in the woods. He returned to work the next day.

38. **Guggenheim fellowship to Berlin**: Muller went to the Kaiser Wilhelm Institute in Berlin on a Guggenheim Fellowship and worked on gene expression of recessive genes in heterozygotes and the detection of semi-lethal genes. Hitler came to power and Muller accepted an invitation to come to the USSR.

39. **Attack on Kaiser Wilhelm Institute**: After the Book Burning Nazi rally in Berlin, a raid was staged by Storm Troopers. Muller escaped and accepted an invitation by Vavilov to go to the USSR in 1933.

40. **To Leningrad and Moscow**: Muller worked on position effect and studied how genes could be switched on or off by moving them to different regions of a chromosome. He also analyzed gene size by using the scute series of genes, inducing many of them as inversions or translocations with one of the breaks near the y-ac-sc region of the X chromosome.

41. **Friendship with Vavilov**: Vavilov was the Head of Agriculture in the USSR and had studied the centers of origin

of domesticated plants. He provided the funds for Muller's research first in Leningrad and then to larger quarters in Moscow.

42. **Hopes for Russian eugenics**: With Levitt, Muller consulted on twin studies at the Medico-Genetic Institute in Moscow. Muller also wrote a book, *Out of the Night* that Levitt suggested could be sent to Stalin.

43. **Lysenkoism in the USSR**: From 1936 on genetics was under attack by Lysenko, a lamarckist agronomist who rejected Mendelism and the chromosome theory of heredity. He believed all traits could be changed by environmental training. When Muller's two Russian students, Agol and Levitt, were arrested and executed, Muller left the USSR.

44. **Volunteer in Spanish Civil War**: The only safe way was by enlisting to fight in the Spanish Civil War and Muller worked with Norman Bethune and the Canadian Blood Unit. After the collapse of the Loyalist forces, Muller was helped in a job search by Huxley.

45. **Divorce from Jessie**: Muller's marriage ended in 1936 with a divorce that gave custody to Jessie and Muller had to make trips during the summer to see his son. Jessie had fallen in love with Muller's student, Offerman, who had accompanied them to Leningrad and Moscow.

46. **Research at Edinburgh**: Huxley recommended Muller to Crew at Edinburgh who was happy to add Muller to the faculty. Muller again built a productive laboratory with several international students, including Pontecorvo and Ray Chaudhuri, and postdoctoral student Charlotte Auerbach.

47. **Marriage to Thea Kantorowicz**: Also, at Edinburgh was Thea Kantorowicz a refugee from Germany via Turkey after

she got her father out of a concentration camp. Her father was head of the Dental School in Bonn.

48. **WWII forces a return to US**: After WWII began Muller had to return to the US. He got a job at Amherst for the duration of the war. When the war ended, he had to find a new position.

49. **Atomic bomb shifts public concerns**: In 1945 the war ended after two atomic bombs were dropped on Japanese cities. Muller recognized that the exposed population had received severe genetic damage, including radiation sickness, near the hypocenter and less catastrophic exposure up to a mile from the center of the explosion.

50. **IU recruits Muller**: Muller was recruited to Indiana University by Fernandus Payne, a student of both Morgan and Wilson. Muller's addition was a happy one for IU because he invigorated the Zoology Department and obtained funding from both the University and Rockefeller Foundation to support his research.

51. **Nobel Prize**: In October 1945, Muller received word that he had won the Nobel Prize for Medicine and Physiology. Muller enjoyed working there until his retirement in 1964.

52. **Cold War radiation controversy**: Muller was a major critic of government policies that exposed the population to ionizing radiation in medicine, industry, and the military use of radiation. He served on national and international committees to protect the public with regulations that offered safety to workers in these fields and to the public exposed to radiation.

53. **Lysenkoism as a worldwide concern**: Also, of concern to Muller, was the attack on genetics in the USSR and the Iron Curtain countries.

54. **Eugenics revived**: In 1957 Muller tried to revive positive eugenics by calling it germinal choice but had little success in getting eugenic considerations to be part of the assortative mating that takes place in marriages.

55. **Teaching and research at IU**: Muller again built up a department with many students (myself included) enjoying his mentoring for our dissertation research.

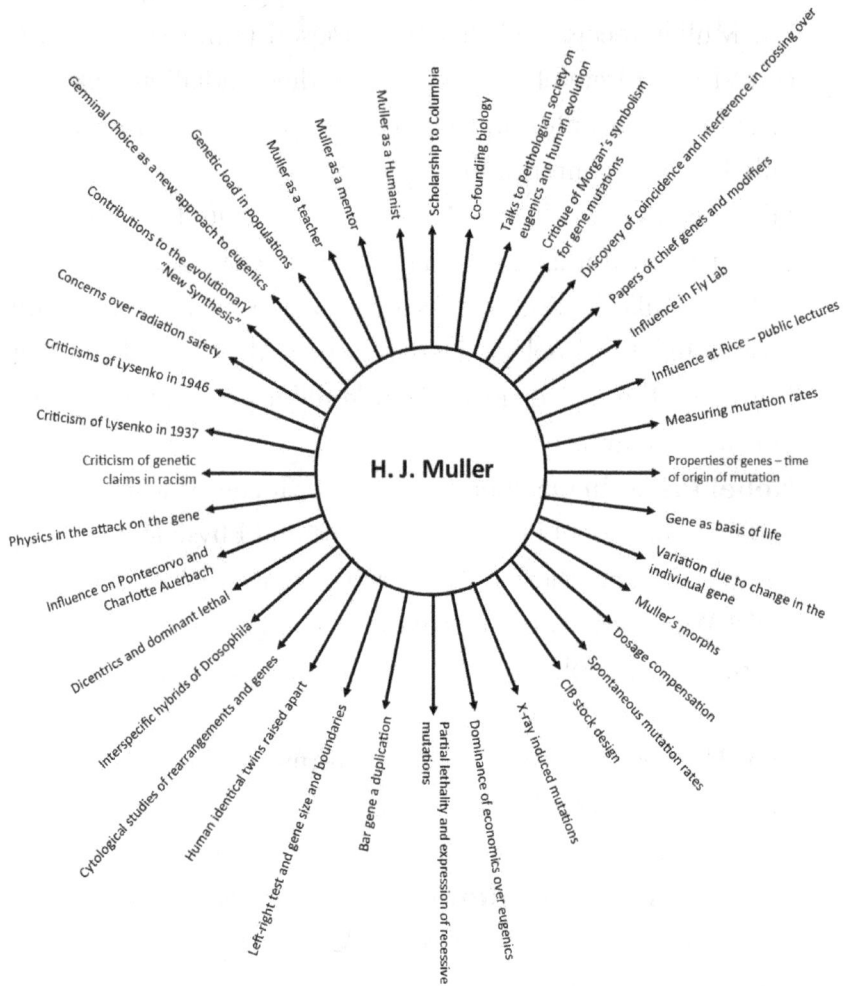

List of Outputs by Muller

1. **Scholarship to Columbia**: Muller appreciated the scholarship to Columbia and lived up to expectations, graduating Phi Beta Kappa.

2. **Co-founding Biology Club**: Columbia's Biology Club was a major factor in forming the fly lab because it brought insiders and outsiders from Morgan's laboratory to discuss genetics.

3. **Talks to Peithologian Society on eugenics and human evolution**: Muller got to know Henle in the Peithologian Society who later became President of Vanguard Press. He was happy to publish Muller's *Out of the Night* in 1936. The book was also published in French and in the United Kingdom.

4. **Critique of Morgan's symbolism for gene mutations**: Morgan had a cumbersome notation that tried to symbolize the inferred functions of mutations rather than their appearances. Muller argued for and convinced geneticists to use the notation w = white eyes, w^+ = red eyes, w^e = eosin eye color, w^i = ivory, etc. For dominant mutations the convention would be a capital letter as in B (bar eyes) Cy (Curly wings), T (truncate) or dp^T (dumpy-truncate), dp^v = (dumpy-vortex), dp^o = (dumpy-oblique), etc.

5. **Discovery of coincidence and interference in crossing over**: The first maps of Sturtevant's were not as precise as the maps drawn two or three years later, which often required numerous crosses of genes with neighboring genes. Muller showed the discrepancies were due to the distance between two genes. Those close together rarely had two exchanges between them. Those far apart had these more frequently and this would register as a non-crossover. Muller provided

the mathematics to predict the distances between two genes when corrected for these probable double crossover events.

6. **Papers of chief genes and modifiers**: Muller studied Truncate wings and Beaded wings, both dominant mutations that were also, as he showed, recessive lethal. He isolated intensifiers and diminishers of expression of the mutant gene or its normal allele.

7. **Influence in Fly Lab**: Muller was a theorizer who saw relations among isolated data and often suggested experiment to test the various theories he and the other fly lab students came up with. Not all his suggestions when proven correct were acknowledged. This led to Muller's resentment and a reaction to it giving Müller the reputation of having a "priority complex."

8. **Influence at Rice — public lectures**: Muller believed science has profound effects on society and he felt he, and his students, should speak out on public issues as much as time and circumstance would provide.

9. **Measuring mutation rates**: Muller and Altenburg carried out experiments to determine the first spontaneous mutation rates in fruit flies.

10. **Properties of genes — time of origin of mutation**: Muller showed how mosaicism could be seen or detected by experimental analysis of spontaneously arising or (about six years later) induced mutations. This included mutations in the germ line or just in the soma or both.

11. **Gene as basis of life**: Muller extended the cell doctrine (all cells arise from preexisting genes) and applied it to the Bar mutation which he showed (with **Alexandra Prokofeyeva Belkovsky**) was a duplication visible in salivary gland chromosomes of fruit fly larvae. He argued that all genes arise

from pre-existing genes and this explained genetic repeats and the number doublets seen in these giant chromosomes.

12. **Variation due to change in the individual gene**: Muller's clarification of the term mutation limited it to a change in the individual gene. Before that mutation was used for any variation however it arose. Calling other changes position effects, aneuploids, or chromosomal rearrangements gave a more precise understanding of the events involved in their formation and their roles in evolution.

13. **Muller's morphs**: Muller classified mutations of the gene as amorphs (complete loss of function such as w — white eyes), hypomorphs (less than full expression of the normal gene as in apricot = w^a), hypermorphs such as e = ebony body color where there is more pigmentation of the body color than normal, and neomorphs in which a new trait is expressed by a mutated gene as in B (for Bar) where reduction in eye size arises from a duplication of the gene.

14. **Dosage compensation**: Muller also discovered while at Texas, that the allele eosin in the white eye series, is not dosage compensated. This applies to X-linked genes because females have two X chromosomes and males have only one X. The male Y has virtually none of the genes found on the X chromosome. Thus, in a female the genotype w^a/w^a = white apricot is the same intensity as male apricot eyes w^a/Y. It is dosage compensated. The gene eosin in females w^e/w^e is darker than the eosin in males w^e/Y. The gene eosin is not dosage-compensated. Dosage compensation was later found in humans, but it arises from inactivation of one of the X chromosomes in the female.

15. **Spontaneous mutation rates**: In fruit flies Muller and Altenburg demonstrated that the spontaneous frequency of mutations is one X-linked lethal per thousand X chromosomes

tested. Mutation rates can vary depending on other genes regulating metabolic events in the cells.

16. **ClB stock design**: Muller X-linked lethal test (modified as the Basc or Muller-5 stock in different countries) played a huge role in radiation genetics studies. It detects mutations by the absence of an expected class of males and thus each vial can be a sampling of just one female carrying (or not carrying) an induced or spontaneously arising gene mutation.

17. **X-ray induced mutations**: Muller's X-ray approach was used on dozens of species and was applied to fungi to produce mutations boosting the production of penicillin which made its use widespread around the world.

18. **Editing *the Spark***: Muller's early communist views were expressed in *the Spark*. The articles included charges that wars are brought about by munitions makers, black people were treated as inferior especially in southern states, the Depression was caused by capitalist greed leading to a market collapse, and that education should be free through all stages of higher education. Today most of these items are considered liberal and not communist except by far-right extremists.

19. **Dominance of economics over eugenics**: Muller's 1932 speech at the third International Congress of Eugenics, denounced the American eugenics movement for its racism, sexism, and questionable use of social status as a genetically imposed condition. It hastened its rejection by geneticists and after World War II and Nazi eugenic excesses, virtually killed the American Eugenics movement.

20. **Partial lethality and expression of recessive mutations**: Müller worked in Berlin and Russia on this project and isolated recessive lethal or visible mutations with partial dominance, demonstrating that even heterozygous recessive mutations have detectable expression.

21. **Bar gene a duplication**: Muller used the bar case to interpret the evolution of life from the first replicating molecules to the formation of genes and through crossover errors of pairing leading to duplications of genes that underwent differential mutational fates leading to new gene functions.

22. **Left-right test and gene size and boundaries**: The left-right test of the y-ac-sc region of the X chromosome demonstrated to Muller that genes had inert chromosomal material (then called heterochromatin) and gene size could be obtained from crossovers between X chromosome carrying different rearrangements with a common or nearly common breakage point.

23. **Human identical twins raised apart**: In 1925 Muller studied identical twins raised apart (both with high IQ scores) and he believed such studies should be done with a careful analysis of traits by skilled psychologists who would control for environmental factors that could make or cause the traits seen.

24. **Cytological studies of rearrangements and genes**: Muller designed stocks to pick up translocations, inversions, and ring chromosomes.

25. **Interspecific hybrids of Drosophila**: Muller showed that most of the traits expressed in hybrids of *D. melanogaster* and *D. simulans* could be teased apart showing which expressed *melanogaster* genes, which expressed *simulans* genes, and which led to new or absent functions. It wasn't a different cytoplasm; it was a different mixing of long separated genes that were involved in the hybrid's overall fertility or viability or interspecific expression.

26. **Dicentrics and dominant lethals**: Muller and Pontecorvo showed that X-rays induced breaks that led to dicentric chromosomes that led to what McClintock independently called the breakage-fusion-bridge cycle leading to cell death. Muller

applied this to the Hiroshima and Nagasaki atomic bomb survivors and claims radiation sickness involves massive production of aneucentric chromosomes and fragments leading to cell death in normally dividing tissues.

27. **Influence on Pontecorvo and Charlotte Auerbach**: Pontecorvo had fled Italy when Hitler and Mussolini agreed to policies limiting rights of Jews. Auerbach also fled Germany because of persecution against Jews. Each established, under Muller, brilliant experiments that gave them lifelong careers in Scotland. Auerbach founded the field of chemical mutagenesis. Pontecorvo studied gene structure after his work on dicentric chromosomes and cell death.

28. **Physics in the attack on the gene**: Muller as early as 1936 hoped physics would be applied to the study of the gene to reveal its unique capacity to replicate its variations. He predicted the use of X-ray diffraction to the gene would someday reveal its structure.

29. **Criticism of genetic claims in racism**: Muller quickly condemned papers and books that assigned racial inferiority or superiority to humans. He felt that such claims lacked experimental evidence or relied on a circular argument that the inferior groups were victims of their genetic inadequacies.

30. **Criticism of Lysenko in 1937**: After Muller left the USSR, he kept quiet about Lysenkoism until 1946 because he wanted to protect the lives and careers of his friends there. It was after the war ended that Lysenko launched a major attack on genetics. Muller learned former friends and colleagues were arrested, demoted, or killed. He denounced the Lysenko movement at the 1946 International Congress of Genetics in Stockholm where he was President of the Congress.

31. **Concerns over radiation safety**: Muller immediately gained international attention when he received the Nobel Prize in 1946 for his work in founding the field of radiation biology. Physicians did not want patients to be scared off from taking diagnostic X-rays. Industry did not want added protections (lead shielding) for workers exposed to relatively low doses of radiation. The military did not want a ban on nuclear weapons. This made Muller controversial although throughout the Cold War he favored bilateral and not unilateral rejection of atomic bomb testing.

32. **Contributions to the evolutionary "New Synthesis"**: Muller was a contributor to the new synthesis that added classical genetics to evolution and Muller showed that Darwinian selection of gradual increases or decreases in function was possible through his study of lethal, semi-lethal, partial dominance of recessive mutations, and the existence of genetic modifiers.

33. **Germinal Choice as a new approach to eugenics**: In 1957 Muller revived his hopes for positive eugenics and hoped that calling it germinal choice would make the proposal acceptable. It fell on deaf ears or put Muller, in the mind of many of his critics, as trying to revive a discredited eugenics movement.

34. **Genetic load in populations**: Muller argued that it is not just isolated mutations that cause damage, but even the partial dominance of recessive lethal was enough in the long run to have detrimental effects on survival of individuals in a population. He called this genetic load, and it was one of the controversial arguments used in the Cold War debates on atomic testing.

35. **Muller as a teacher**: Müller was not a polished lecturer. He thought while he was lecturing and often came up with

new ideas and connections that excited him as much as they excited the class. He also used a historical approach to each topic discussed in his courses so the class could see that ideas evolve.

36. **Muller as a mentor**: Müller read every sentence of our articles, progress reports, or dissertation drafts. He did so in a one-hour session we each had during the week in his office and these were exhilarating because we learned to fight to preserve or defend our interpretations and often ended up doing more experiments to test the issues raised in these discussions. It sharpened our minds as scientists.

37. **Muller as a Humanist**: Muller believed in ideals to live by. These came from human experience and not gifts of gods. He believed humans had enormous capacities to use their knowledge for the good of society and the goal of Humanism was to constantly seek the good that could make life healthier, safer, and happier. He rejected any supernatural reliance guiding us or serving as a distant goal.

16 Comments for Muller's Outputs

Muller early showed a love for science and its applications to benefit humanity. Partly this might have been his exposure to Unitarian outlooks in his youth. It carried over to the Humanism that he embraced in later life. Muller is best known for his work on radiation genetics which netted him a Nobel Prize. He is known among geneticists for his studies of the gene — its size, number, physiological functions, and role in evolution. Muller promoted and marshaled evidence for "the gene as the basis of life." He made substantial contributions as a graduate student to the theory of the gene. He provided much of the genetic evidence for Darwinian evolution through his analysis of the Beaded and Truncate genes which showed how modifier genes accumulated to stabilize a trait.

In his 1926 paper, "The gene as the basis of life," Muller attributed to the gene the capacity to replicate its errors. This was an essential property for the evolution of life. Instead of a constant crystal, like table salt, which is always Na+ Cl-, the gene had a crystal-like structure allowing it to replicate and a chemical diversity in gene mutations that did not simultaneously disrupt its replication. It was not until 1953 that DNA was interpreted by Watson and Crick as being an aperiodic crustal and its double helix structure associated with the specific pairing of purine and pyrimidine held by hydrogen bonding. For Muller this was the molecular basis of gene structure and functions. Watson attributed his interest in gene structure as an outcome of taking Muller's courses at Indiana

University. Muller's influence was also felt in the "new synthesis" of the 1940s that brought together biological geography, systematics, comparative anatomy, paleontology, population genetics, cytogenetics, and natural selection.

Muller embraced socialism, but in its Bolshevik form, it was abused by totalitarian personalities like Lenin and Stalin. Muller became a foe of the misuses of genetics for racism, nationalism, and discrimination, which falsely attributed a genetic basis for despotic goals. He spoke out against such misuses of genetics in his articles and public speeches. He did so at considerable risk and had to abandon jobs when the politics against him endangered both his career and personal safety. After receiving his Nobel Prize he became a leading proponent of radiation safety in what was then called the Atomic Age. He also was the leading critic of the destruction of genetics in the USSR by T. D. Lysenko and his supporters.

17 Evaluating the Outputs of Scientific Careers

In my own life I early assumed the role of teacher's pet because I did well in class and came up with original questions. I asked one teacher in junior high school if the idea of 360 degrees in a circle was roughly based on the 365 days in a year. It was, and epicycles were introduced in the Ptolemaic view of the universe to keep the circularity model in describing the various components of the universe. On another occasion, in high school, a teacher was trying to remember the name of a religious leader who opposed Hitler's policies and who was executed for his continued criticism. "Was it Pastor Niemuller?" I asked. It was, and I learned that I had what I call a "fly paper memory." If odd facts are interesting, I somehow can retrieve them. I had remembered hearing his story during WWII on a radio program called "This is the enemy." My writing got its start from doing essays in my high school English classes. I would sometimes submit two different essays on a topic if I got into reading about a topic. I got feedback from colleagues that my writing style was very appealing to them. I began writing books after I established my career as a geneticist with an active graduate program that led to six students getting their PhDs. My first published book was *The Gene: A Critical History* written during my first sabbatical leave at UCLA. Muller was still alive, and he liked my account of how the gene developed from vague ideas of hereditary units in the 1860s to the molecular basis of heredity. I asked him if I could write his biography. I did interviews with his colleagues and family and used a teaching award (the Harbison Prize) to fund a trip to do interviews in Europe. The bulk of my reading came after Muller died and I worked with Thea Muller on the collection she donated to the Lilly Library at Indiana University. I called that book *Genes, Radiation, and Society: The Life and Work of H. J. Muller.* I have had 13 books published and continue to write in my old age. Most are scholarly books because I want them to live on in university libraries and reveal to future generations how science historians and philosophers approached their fields of interest. Among these books I would single out *The Unfit: A History of a Bad Idea* for its history of the ties of degeneracy theory to the emergence of the eugenics movement as my most exciting topic to research, especially at the Kinsey Institute for Sex Research Library at IU. It has numerous books on masturbation, a craze since the early 1700s associated with degeneracy theory, a precursor of eugenics. I also consider *How Scientific Progress Occurs: Incrementalism and the Life Sciences* as a better documented history of how the life sciences work than does T. J. Kuhn's theory of paradigm shifts. This short work on Intellectual Pedigrees provides ample information on the importance of mentoring in teaching students. I identify myself as a generalist. In this respect I am more like Julian Huxley or J. B. S. Haldane than I am like Muller, whose commitment to research was lifelong.

In Muller's life, he focused on genetics from his undergraduate days to his death at age 77. He was a materialist who championed reductionist approaches to research, and he relied on both theory and experimentation. He never thought of his science as a game. It had both theoretical implications and applications to human welfare. He saw the gene as pervasive in the life sciences. His courses on "mutation and the gene," "radiation genetics," and "evolution" were powerful influences on his students. Muller lived at a time when world upheavals were numerous. He recognized that genetics was a threat to totalitarian states who misused genetics to support racism and genocide. He also saw genetics falsely called a Capitalist pseudoscience that should be replaced with a politically created Lysenkoism that argued for a total environmental determinism in biology. In the USSR that led to the arrest, execution, demotion, or exiling of critics of Lysenko's state-backed brand of the life sciences. During the Cold War, Muller's concerns about radiation safety were attacked as Communist propaganda in the US while the USSR was attacking Muller for promoting Mendelian genetics to justify Capitalism and Social Darwinism.

The circles of inputs or outputs are lengthy for me and for Muller. I had the advantage of doing Muller's biography and thus have more detailed knowledge of those inputs and outputs, almost (but never as fully) as my own. The more distant the scientist we choose to study for inputs and outputs the less there will be available of surviving documents to use. What applies to the fossil record for the story of evolving species, also apples to our scientific lives. There are only a few documents written by Copernicus that have survived. The availability of printing made his *Commentariolis* and his posthumous book on the solar system the major sources for knowledge of his work in astronomy. I hope other scientists will try to make their own circles of inputs and outputs. The richness of dozens or hundreds of such living scientists participating in this exercise will be a gift to the study of how science works and what is universal and what is peripheral in a scientific career.

18 Reflections on Constructing Intellectual Pedigrees

I first constructed an intellectual pedigree in the 1960s. I used Alfred Sturtevant's pedigrees of geneticists in his *History of Genetics* (1965) as a starter. I got to **Döllinger** and thought that was probably the last connection I would find. When I did this more extensive pedigree of my past mentors on January 14 and 15 of 2017, I was overwhelmed by what I found. I never expected to find **Darwin, Newton, Malpighi** and **Galileo** as direct ancestors of this intellectual pathway. Is it possible that most geneticists would find a connection leading to Galileo? Surely there are those who never attended college who became scientists and have no direct connection with an academic mentor. It is also possible that biologists in different specialties will be able to hook up to the Galileo line as often as it does among geneticists.

I used Wikipedia for the photos and biographical entries. For the US 20th century and late 19th century I used the *Biographical Memoirs of the National Academy of Sciences*.

I hope that this approach will be used by others to explore how intellectual influences travel from country to country. It was **Agassiz** who brought the European lineage to North America in this pedigree. It was **Döllinger** who brought the Italian lineage to Germany and France. Some scientists had two rather than one mentor that led to their field of research (e.g., Morgan and Wilson influencing Muller). I also suspect this will be more difficult in the humanities where one-on-one mentorship is rare for graduate students. Following this lineage also shows how knowledge is bundled. In the Renaissance students like Copernicus could study canon law, astrology, mathematics, and medicine, often going from city to city to find the specialists they desired. In the 19th century after the **Humboldt** brothers established the PhD research university, European scholars frequently went to three or more universities to learn cognate fields. Today the concentration is more exclusive, and students must first finish a PhD at one institution in order to move on to a post-doctoral sponsorship.

I have done rough intellectual pedigrees (see Appendix) for **Ralph Cleland** who goes back to **Hugo von Mohl**. **Barbara McClintock** goes back to **Carl Linnaeus**. **Seymour Benzer** in one branch goes back to **Friedrich Wöhler** and in a second branch goes to **Andrea Ettinghausen** (**Gregor Mendel's** University of Vienna mathematics teacher). **Wilhelm Johannsen** in one branch goes back to **Theodore de Saussure**.

There are branches where a scientist had two or more mentors. Thus, **Morgan** and **Wilson** were major contributors to **Muller's** outlook. Similarly, **Louis Agassiz** had two major mentors, **Georges Cuvier** and **Alexander von Humboldt**. I followed both branches for Muller's lineage but for Agassiz I have only followed one line, Cuvier's. A rough estimate from what I could find is that **Alexander von Humboldt's** intellectual pedigree would go back to **Immanuel Kant** and **Gottfried Leibnitz** (father of the more famed philosopher son).

In human pedigrees little is known of our biological ancestors if we are not descended from royalty or social elites (like many of those who are descendants of the *Mayflower*). Virtually none of their writings or thoughts persist. But in intellectual pedigrees, ideas matter and so do scientific discoveries which become enshrined in the scientific fields where they arose. The scientific findings and observations get published. Our ancestors' working lives rarely get published and except for rites of passage (birth, marriage, and death) no written records are available for most of our ancestors who lived over 100 years ago.

Missing from this intellectual pedigree are the many contributions of techniques, equipment, and unpublished (and unacknowledged) conversations that are part of how science works. I hope the input and output assessments will be explored by those in history and philosophy of science that seek a more complete picture of how science works. It may well require another generation and advances in the molecular biology of mental activity before we can infer how these numerous incoming experiences, conversations, and readings shape a scientist's personality and habits.

Of interest to historians of science is the relation of teachers to their students. Mentoring is intense because there are lots of discussions and a professor may see a few students several hours a week, in contrast to giving lectures to hundreds of students per week. **Whisson** at Cambridge coached 206 students in a nearly 20-year period from 1744–1763. The vast number of these did not enter academic careers or their mentored experience did not lead to a successful career measured by noted descendants. Missing from this intellectual pedigree are the many contributions of techniques, equipment, and unpublished (and unacknowledged) conversations that are part of how science works.

Also, of interest to historians of science, are the number of universities and the number of faculty in natural sciences from the dawn of European Universities (starting in 1088) to the modern dissertation-based PhD in 1810). The total faculty in a medieval or renaissance university may have been a dozen or less with only one or two teaching the natural sciences. This may explain why so many end up with Galileo if the pedigree is traced to the 1500s.

Appendix A
Using Non-illustrated
Vertical Intellectual
Pedigrees

In this format names are given, and the pedigree is read from the most recent downward to the most distant mentoring relationship. Unless otherwise discussed, it is assumed that each person named held some sort of academic degree. In my pedigree there are 16 generations to Tartaglia, and all had a university affiliation. In the line of descent from Newell Martin to Isaac Newton, Charles Darwin would be the only member who did not have a university setting for his scientific work. Darwin used his own wealth to support himself and worked at home after returning from the *Beagle* voyage. I included Morris Cohen in this vertical pedigree although his contribution was in the humanities. I included him in discussing my own intellectual career because writing books and relating science to the liberal arts has been a major part of my career. There are many ways to prepare intellectual pedigrees and a legend is helpful to readers. Some prefer to construct an academic pedigree from the oldest ancestor on top and most recent scientist at the bottom. The analogy is one of a royal family beginning with the oldest member on top. The tree model assumes the oldest are at the root level and the higher branches are the most recent.

When fleshed out, as I have done in this book, the intellectual pedigree provides an opportunity to study the history of a field, the social circumstances in which academic science was performed, and the way scientific fields moved across Europe and into the New World. I hope many of the scientists and scholars reading this book will prepare their own intellectual pedigrees and make these available to appropriate websites. I do not doubt that scholarly intellectual pedigrees in many fields can be constructed in the major Asian civilizations of India, China, and Japan. They can also be constructed in the Islamic countries. In Japan, genetics and modern experimental biology were brought to their universities by **Charles Otis Whitman** (1842–1910) who later taught at the University of Chicago and was first Director of the Marine Biology Laboratory at Woods Hole. Many of the lost works of Greek and Latin scholars were retrieved and translated from the establishment of Islamic centers in Spain and through the Mediterranean ports where works from Asia and the Middle East found a ready market of scholars from Italy, Spain, and Portugal throughout most of the Middle Ages. I have provided brief accounts of the lives of predecessor mentors. They can be expanded (as in pages 10–23) if you prepare your own intellectual pedigree.

My preference for a vertical (present to past in a downward path) presentation is based on Newton's famed quote of standing on the shoulders of giants. I like that image of what we owe to the past.

Some issues are difficult to assess. A reading of a book (as in many molecular biologists who read Schrödinger's *What is Life?*) may have an indirect influence on shaping a career because its impact is profound. Similarly, not everyone who attends a lecture that shifts a career makes note of that event. Darwin's *Origin of Species* played a similar role in the careers of many nineteenth century scientists. Some geneticists did their PhDs with uninspiring mentors but later through reading and attending meetings or having colleagues at their university appointment were fortunate to be shifted into becoming outstanding geneticists. Mendel had no direct heirs to his findings in the 1860s and it was the 1900 rediscovery that made him leapfrog to iconic status. In this more complex sense of how science works, these academic pedigrees are guides rather than proven pathways to the mentoring process most geneticists have experienced.

My estimate for your academic pedigree is that you have a 67 percent chance that you will connect to Newton or Galileo or both. This comes from the tabulation in Table 1. In this Appendix, 16 go to Newton, 15 of the lineages go to Galileo, and 15 go to neither Galileo nor Newton.

Table 1   Frequency of intellectual pedigrees associated with Newton or Galileo

to Newton	to Galileo	to neither Newton nor Galileo
Carlson	Carlson	Brenner
Muller	Muller	Correns
Auerbach	Auerbach	Bateson
Castle	Benzer	Luria
Davenport	Castle	McClintock
Sagan	Davenport	Mendel
De Vries	Lederberg	Pauling
Galton	Sagan	Haldane
Morgan	Morgan	Lederberg
Kimura	Kimura	Avery
Beadle	Beadle	Delbruck
Schrodinger	Watson	Crick
Watson	McKusick	Borlaug
McKusick	Wright	Swaminathan
Fisher	Beadle	Goldschmidt
Beadle		
16	15	15

Note: 12 have both Newton and Galileo

Appendix B
Reference Intellectual Pedigree for Hermann Joseph Muller

[for paths to **NEWTON** and/or **GALILEO**]

Hermann Joseph Muller [1890–1967]

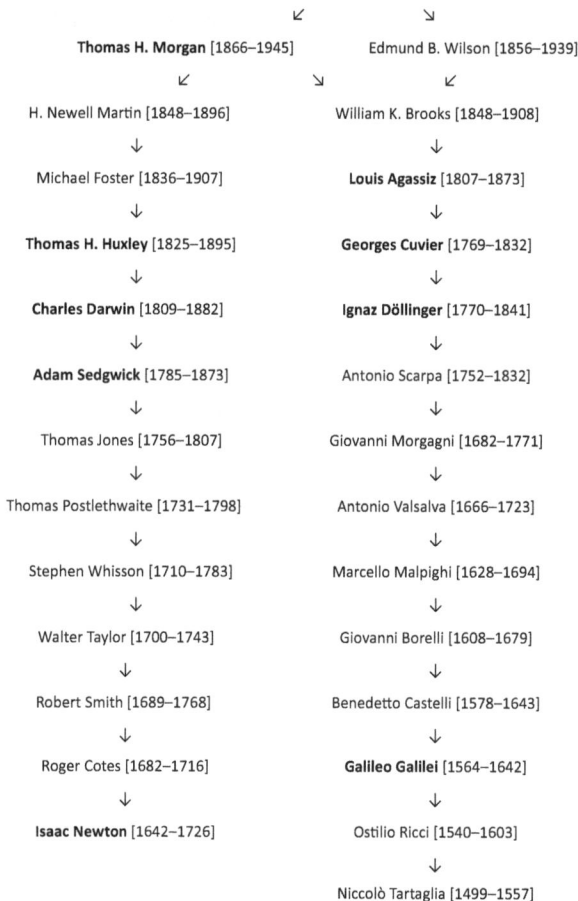

↙   ↘

Thomas H. Morgan [1866–1945]   Edmund B. Wilson [1856–1939]

↙   ↘   ↙

H. Newell Martin [1848–1896]   William K. Brooks [1848–1908]

↓   ↓

Michael Foster [1836–1907]   **Louis Agassiz** [1807–1873]

↓   ↓

Thomas H. Huxley [1825–1895]   **Georges Cuvier** [1769–1832]

↓   ↓

Charles Darwin [1809–1882]   **Ignaz Döllinger** [1770–1841]

↓   ↓

Adam Sedgwick [1785–1873]   Antonio Scarpa [1752–1832]

↓   ↓

Thomas Jones [1756–1807]   Giovanni Morgagni [1682–1771]

↓   ↓

Thomas Postlethwaite [1731–1798]   Antonio Valsalva [1666–1723]

↓   ↓

Stephen Whisson [1710–1783]   Marcello Malpighi [1628–1694]

↓   ↓

Walter Taylor [1700–1743]   Giovanni Borelli [1608–1679]

↓   ↓

Robert Smith [1689–1768]   Benedetto Castelli [1578–1643]

↓   ↓

Roger Cotes [1682–1716]   **Galileo Galilei** [1564–1642]

↓   ↓

Isaac Newton [1642–1726]   Ostilio Ricci [1540–1603]

↓

Niccolò Tartaglia [1499–1557]

Commentary on the Muller Reference Pedigree

This is the first completed intellectual pedigree I prepared (I omit myself in this reference pedigree). It is in the linear, present to past (top to bottom), direction with arrows pointing to mentors. It contains two major branches, because Morgan owed as much to **Brooks** as to **Martin**. The Brooks line goes from **Agassiz** to **Galileo**. The Martin line goes to **Darwin** and **Newton**. The biographical paragraph for each of these individuals is given in the illustrated version. That is the form that takes several pages (pp 20–32) to read. The short form presented in this Appendix is followed in each subject by a shorter account of the individual mentors. Because the Galileo or Newton lineage is found in many of the scientists whose pedigree is shown, I have used the bold-faced member of a lineage as a terminus in these pedigrees. I follow this with a reference to a "see pedigree" to the Galileo, Newton, or other specified line.

Appendix C
Intellectual Pedigrees

Charlotte Auerbach [1899–1994]

↙   ↓   ↘

Otto Mangold [1891–1962]   Francis A. E. Crew [1886–1973]   H. J. Muller [1890–1967]

↓   ↓   see **Muller** pedigree

Hans Spemann [1869–1941]   Arthur D. Darbishire [1879–1915]

↓   ↓

Carl Gegenbaur [1826–1903]   Walter F. R. Weldon [1860–1906]

↓   ↓   ↘

Albert Kölliker [1817–1905]   Francis Balfour [1851–1882]   E. Ray Lancaster [1847–1929]

↓   ↓   ↙

Jacob Henle [1809–1885]   Michael Foster [1836–1882]

↓   ↓

Johannes Müller [1801–1858]   **Thomas Huxley** [1825–1895]

↓   ↓

August Mayer [1787–1865]   to **Newton**

↓   in **Muller** pedigree

Karl Kielmeyer [1765–1844]

↓

Georges Cuvier [1769–1832]

↓

to **Galileo**

in **Muller** pedigree

Pedigree of Charlotte Auerbach

Charlotte Auerbach (1899–1994) was born in Krefeld, Germany, and died in Edinburgh, Scotland. Her father was a chemist; and her grandfather was an anatomist. She attended college in Würzburg, Freiburg, and Berlin. At the Kaiser Wilhelm Institute she started her doctorate work in embryology with **Otto Mangold**, but had to leave when the Nazis assumed political power. She moved to Edinburgh and received a PhD under **F. A. E. Crew**. She did her postdoctoral work with **H. J. Muller** when he was at Edinburgh (1937–1940). Auerbach used mustard gas to induce mutations in fruit flies and characterized the differences between chemical mutagenesis and radiation-induced mutagenesis.

Branch 1: From Mangold

Otto Mangold (1891–1962) was born in Avenstein, Germany, and died in Heiligenberg, Germany. He was an embryologist who worked with Spemann on the induction of embryonic change using experimental embryology to identify "organizers." Spemann was an authoritarian Professor who embraced Nazism when it took over the government and he supported racial laws enacted by the Nazis. Auerbach, who was Jewish, had to leave.

Hans Spemann (1869–1941) was born in Stuttgart, Germany, and died in Freiberg, Germany. He won a Nobel Prize for his work in experimental embryology using microdissection to detect differentiated stages of embryonic cells. He showed there were regions of tissue that released diffusible substances that acted as organizers of organ formation.

Carl Gegenbaur (1826–1903) was born in Würzburg, Germany, and died in Heidelberg, Germany. He was a histologist studying normal and pathological tissues and their cells. He had several mentors including **Kölliker, Virchow, Müller,** and **von Leydig.**

Albert Kölliker (1817–1905) was born in Zurich, Switzerland, and died in Germany. He named protoplasm (the cell's living material) and cytoplasm (the material surrounding the nucleus and bounded by the cell membrane).

Jakob Henle (1809–1885) was born in Furth, Germany, and died in Göttingen, Germany. He studied the anatomy of the adrenal gland and identified cells of the adrenal cortex were distinct from those of the adrenal medulla.

Johannes Müller (1801–1858) was born in Koblenz, Germany, and died in Berlin. He identified the embryonic ducts associated with the reproductive system. The Mullerian ducts are associated with the formation of the uterus and the fallopian tubes.

August F. J. Mayer (1787–1865) was born in Schwabia and died in Bonn, Germany. He was an anatomist and first to describe the anatomy of Neanderthal skeleton found in 1856.

Karl Friedrich Kielmeyer (1765–1844) was born in Bebenhausen, Germany, and died in Stuttgart, Germany. He was a chemist and botanist. He used comparative anatomy to suggest that body plans were evidence of a past evolution. He believed that all living things were endowed with a vital spirit.

Georges Cuvier (1769–1832) was born in Montbéliard (later French) and died in Paris, France. He was a Protestant and largely an autodidact. Cuvier studied comparative anatomy and believed that the geological strata in which fossils were found suggested a series of major catastrophes that wiped out those forms of life and they were replaced by new creations of life different from the prior and underlying rock formation. He rejected the evolutionary model of Lamarck.

Auerbach's second branch

Francis Albert Eley Crew (1886–1973) was born in Tipton, England, and died in Edinburgh, Scotland. He was a geneticist who studied the genetics of reproduction in animals, especially mammals and birds and interpreted intersex conditions and cryptorchy. He headed the Institute for Animal Genetics at the University of Edinburgh.

Arthur Dukinfield Darbishire (1879–1915) was born in Kensington, England, and died of meningitis in London during World War I. His father was a physician. He received his MA at Oxford and was mentored by **W. F. R. Weldon.** He began Mendelian crosses with mice and became England's first geneticist to hold an academic post with that field's name. He embraced Mendelian genetics when he read Mendel's papers and believed that Mendelism and biometrics were compatible.

Walter Frank Raphael Weldon (1860–1906) was born in London and died at Oxford, England. His father was a chemist. Weldon applied mathematics to populations and heredity and was a founder of the field of biometry with **Karl Pearson**. He was mentored by **Francis Maitland Balfour** and **E. Ray Lankaster**.

Francis Maitland Balfour (1851–1882) was born in Edinburgh and died in Chamonix, Switzerland. Balfour was an embryologist and worked out the urogenital system in fish. He died young while trying to climb Mont Blanc in the Alps.

Michael Foster (1836–1907) was born in Huntington, England, and died in London. He was a physiologist and studied mechanisms of heart beating.

E. Ray Lankaster (1847–1929) was born and died in London. His father was a coroner and his mother was a botanist. He studied the comparative anatomy of horseshoe crabs and arachnids and showed that they were related in their evolution. He also studied parasitism and thought it was another branch of evolution in which degeneracy of organ systems accompanied extreme specialism in their mode of parasitism. Lancaster's work was used by forerunners of the eugenics movement as evidence that social failure in humans was a form of parasitism and degeneracy. Lancaster never married. He was arrested but acquitted of charges of homosexuality. Both **Foster** and **Lankaster** were mentored by **Thomas Henry Huxley**.

Thomas Henry Huxley (1825–1895) was born in London and died in Eastbourne, England. Huxley was an autodidact who became a prominent biologist. He served as a ship's surgeon on a voyage to the South Pacific and his work on the Coelenterates gained him early recognition. He was influenced by **Charles Darwin** and accepted Darwin's theory of evolution by natural selection. He taught at the London School of Mines and was a gifted teacher and essayist promoting the life sciences and their inclusion in the liberal arts.

Auerbach's third branch

Hermann Joseph Muller (1890–1967) (see Muller pedigree).

Oswald Avery [1877–1955]

↙   ↘

Benjamin White   Alphonse Raymond Dochez [1882–1964]

↓   ↓

Max von Gruber [1853–1927]   Eugene Lindsey Opie [1873–1971]

↓   ↓

Carl Nägeli [1817–1891]   William Welch [1850–1934]

↙   ↘

Julius Cohnheim [1839–1884]   Rudolph Virchow [1821–1902]

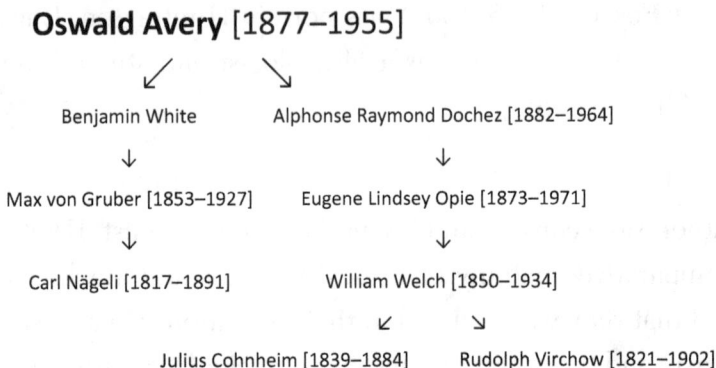

Legend for Pedigree of Oswald Avery

Oswald Avery (1877–1955) was born in Halifax, Nova Scotia, and died in Nashville, Tennessee. His parents moved to New York City when he was 10 years old. He went to Columbia University for his MD degree and chose infectious diseases, especially pneumonia, as his field of study. He had read Frederick Griffith's work on transformation of pneumococcal strain that lacked a capsule that could be made to produce a capsule from the dead extract of infectious pneumococci with capsules. He, Maclyn McCarty, and Collin Macleod in 1943 used preparations of the carbohydrates identified by Avery, preparations of RNA, preparations of proteins, and preparations of DNA from these strains and showed that the DNA of the virulent form would transform non-virulent strains without capsules into virulent strains with capsules. This was the first direct evidence that DNA was the genetic material.

Benjamin White (1879–1938) was born and died in New York City. He received his PhD at Yale and studied in Germany with Max Gruber before returning to join the Hoagland Laboratory in Brooklyn. White devoted his research to bacteriology, working on syphilis, tuberculosis, and other infectious diseases.

Max von Gruber (1853–1927) was born in Vienna, Austria, and died in Berchtesgaden in Germany. His father was a physician. He discovered agglutination, the immune response to injected bacteria or bacterial extracts. This allowed bacteriologists to identify different strains of a bacterial type.

Carl von Nägeli (1817–1891) was born in Zurich, Switzerland, and died in Munich, Germany. He studied botany with A. de Candolle and was stimulated by Schleiden's work on cells to study plant cell morphology. He named xylem and phloem in plant circulation systems and he distinguished the cell wall from the rest of the cell. He inferred hereditary units resided in the cell and called this material idioplasm. He corresponded with Mendel and showed Mendelian ratios were not found in Hieracium, the plant he studied. They exchanged seeds and confirmed each other's findings. This led Mendel to abandon farther studies of peas.

Alphonse Dochez (1882–1964) was born in San Francisco, California, and died in New York City. His family was of Belgian descent. He got his BA and MD at Johns Hopkins Medical School in 1907. He worked on the variety of bacterial types associated with pneumonia. He tried an immune based therapy for treating pneumonia. He was the first to identify the common cold as caused by viral rather than bacterial infection.

Eugene Lindsay Opie (1873–1971) was born in Staunton, Virginia, and died in Bryn Mawr, Pennsylvania. His father was a physician (Obstetrics/Gynecology). He identified the Isles of Langerhans in the pancreas as the probably tissue associated with diabetes.

William Welch (1850–1934) was born in Norfolk, Connecticut, and died in Baltimore, Maryland. He came from a family of physicians and went to Columbia University for his MD. While in Europe he was impressed by their public health programs and research. He brought back a commitment to generate such programs in the US and did so becoming the founder of the first public health school in the US and becoming the first dean of the new Johns Hopkins Medical School. Two of his students went on to win Nobel Prizes (Peyton Rous and George Whipple). He identified the bacterium (*clostridium*) associated with gangrene.

Julius Friedrich Cohnheim (1839–1884) was born in Demmin, Germany, and died in Leipzig, Germany. He received his MD in Berlin and served in the medical corps in Germany's war against Denmark. He demonstrated that the origin of pus was from a migration and overgrowth of white blood cells at the site of infection.

Rudolph Virchow (1821–1902) was born in Swidwin, Prussia (now Poland), and died in Leipzig, Germany. He studied with Johannes P. Müller for his MD and was sent to Silesia to study the cause of high mortality among miners. He blamed low wages and took part in the 1848 Revolution. He was fired and shifted to pathology and showed cancer was a cellular disorder and that all cells arose from pre-existing cells. His idea was a confirmation, or an independent finding shared by Remak and called by Virchow the cell doctrine. He debunked the Aryan movement in Germany and showed that most Germans were neither blond nor Aryan. He was a contrarian in his views of evolution and the germ theory which he both rejected but committed to social applications of medicine to society and a founder of the Public Health Movement in the last half of the 19th century.

William Bateson [1861–1926]

↙	↘

William Keith Brooks [1848–1908]	W. F. Raphael Weldon [1860–1906]
↓	↓
See pedigree to **Galileo**	Francis Balfour [1851–1882]
	↓
	Michael Foster [1836–1907]
	↓
	See **Muller** pedigree **Foster to Newton**

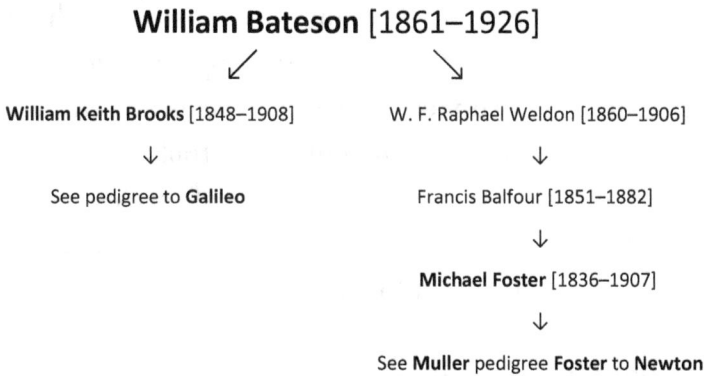

Legend for Intellectual Pedigree of William Bateson

William Bateson (1861–1926) was born in Whitby and died in London. His father was Master of St. John's College at Cambridge. Bateson attended Cambridge for his undergraduate degree where he studied biology, especially embryology with **Francis Balfour** and mathematics with **W. F. Raphael Weldon**. He spent a summer at the Chesapeake Bay summer camp of **William Keith Brooks**. There he used *Balanoglossus*, an acorn worm (hemichordate), for evolutionary studies recommended by Balfour. Brooks told Bateson the most important field to emerge will be heredity which was still searching for a material basis that would be compatible with experimental study. Bateson came back and shifted his attention to variations, writing a book on it after identifying new variations as arising by sudden changes or mutations that he called meristic (duplication of parts) or homeotic (misplaced parts like a leg instead of an antenna in an insect). Bateson rejected Weldon's approach and embraced Mendel's findings confirming Mendelism in animals and extending Mendelism in plants.

W. F. Raphael Weldon (1860–1906) was born in London and died at Oxford. He studied with **E. Ray Lancaster**, an embryologist, in London and then went to Cambridge to study with Francis Balfour. He promoted the biometric approach to evolution (partially through **Francis Galton's** influence). He rejected Bateson's interpretation that discontinuous traits were more significant than continuous ones for the origin of species and a bitter feud broke out between Bateson's school and the biometric school proposed by Weldon and Karl Pearson.

Francis Balfour (1851–1882) was born in Edinburgh and died in Chamonix in the French Alps. He was an embryologist whose work included the formation of the urogenital system in vertebrates, mostly through using fish embryos. He was gaining international recognition at age 31 when he went to the Alps to climb Mont Blanc and he and his guide were killed in an accident.

George Beadle [1903–1989]

✓   ✓   ↓   ↘

Franklin Keim	→ Rollins Emerson	Alfred Sturtevant	Boris Ephrussi	Edward Tatum
[1885-1956]	[1873–1947]	[1891–1970]	[1901–1979]	[1909–1975]
↓	↓	↓		↓
Edward Fred	Edward M. East	T. H. Morgan		Louis Rapkine
[1887–1981]	[1879–1938]	[1866–1945]		[1904–1948]
	↓	↓		↓
	See **E. M. East**	See **H. J. Muller**		Fauré-Fremiet
	pedigree to	pedigree to		[1883–1971]
	Lavoisier	**Newton** and **Galileo**		

Legend for Pedigree of George Beadle

George Beadle (1903–1989) was born in Wahoo, Nebraska, and died in Pomona, California. His parents were farmers. He planned an agricultural career at the University of Nebraska for his BA but his mentor, Franklin Keim advised him to look into genetics at Cornell. He got his PhD with R. Emerson in 1931 and made substantial contributions to maize genetics. He went to Caltech to work with Sturtevant on fruit flies and then went to Paris to study developmental genetics with Boris Ephrussi. They worked out the "one gene–one enzyme" model of biochemical genetics. Beadle extended this to the fungus Neurospora after teaming up with Edward Laurie. The work, that used simple molecules to work out biochemical pathways, earned them a Nobel Prize in medicine. Later in his career, Beadle worked on the evolution of maize from teosinte.

Franklin Keim (1885–1956) was born in Hardy, Nebraska, and died in Washington, D.C. His father was a farmer. Keim studied hybrid wheat and earned his PhD with R. A. Emerson. Keim recommended students with an interest in plant genetics to Emerson's group at Cornell.

Rollins Adams Emerson (1873–1947) was born in Pillar Point, NY, and died in Ithaca, NY. He received his PhD in 1913 from E. M. East and he founded the maize school of genetics which was to plant genetics what the Fly Lab was to Morgan. Emerson's students **G. W. Beadle** and **B. McClintock** went on to win Nobel Prizes.

Edward Murray East (1879–1938) was born in Du Quoin, Illinois, and he died in Boston, Massachusetts. His father was a mechanical engineer. He worked out multiple factor inheritance in maize and devoted most of his career to studying inbreeding and helped develop hybrid corn for commercial use. He was an ardent supporter of the eugenics movement and held racist views and promoted restrictive immigration and compulsory sterilization of those deemed unfit to reproduce.

Alfred Henry Sturtevant (1891–1970) was born in Jacksonville, Illinois, and died in Pasadena, California, In Morgan's laboratory he prepared the first genetic map of the X chromosome of the fruit fly. He did the first comparative genetic study of the chromosomes and genes of related species in fruit flies.

Boris Ephrussi (1901–1979) was born in Moscow and died in Gif-sur-Yvette, France. His father was a chemical engineer. He left Russia after the Revolution and worked in France and went to study with Morgan's group at Caltech where he and Beadle devised experiments to study eye color inheritance in fruit flies, using embryonic eye transplants to work out the biochemical pathway to the two eye color pigments present in flies. He later shifted to yeast genetics and discovered a loss of mitochondria as a cause for tiny colony formation among yeast mutations.

Louis Rapkine (1904–1948) was born in Belarus, Russia, and died in Paris, France. His parents moved to France, then to Montreal, before returning to France. He got his MD at Montreal. Rapkine studied biochemistry in bacteria.

Emmanuel Fauré-Fremiet (1883–1971) was born and died in Paris, France. His father was a noted composer of classical music. At the Sorbonne, Fauré-Fremiet received his PhD. He studied nucleocytoplasmic relations in protozoa and used the electron microscope to identify chromosomes and organelles in protozoa.

Edward Broun Fred (1887–1981) was born in Middleton, Virginia, and died in Madison, Wisconsin. He got his PhD in 1911 in Göttingen and studied bacteria, applying his research to nodules of roots that contained bacteria that helped plants obtain nitrogen compounds.

Edward Lawrie Tatum (1909–1975) was born in Boulder, Colorado, and died in New York City. He obtained a PhD in chemistry in 1934 at the University of Wisconsin. He worked with Beadle to shift Beadle's interests in biochemical pathways from fruit flies to neurospora, using the mold to work out the synthetic pathways for vitamins and other simple molecules present in metabolism.

Seymour Benzer [1921–2007]

Part 1: Through Schrödinger and Lark-Horvitz

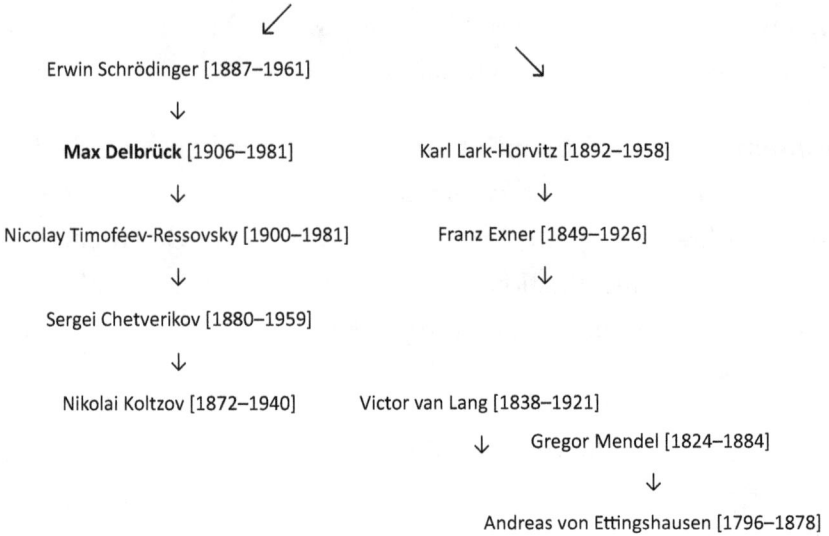

Erwin Schrödinger [1887–1961]

↓

Max Delbrück [1906–1981]   Karl Lark-Horvitz [1892–1958]

↓   ↓

Nicolay Timoféev-Ressovsky [1900–1981]   Franz Exner [1849–1926]

↓   ↓

Sergei Chetverikov [1880–1959]

↓

Nikolai Koltzov [1872–1940]   Victor van Lang [1838–1921]

↓   Gregor Mendel [1824–1884]

↓

Andreas von Ettingshausen [1796–1878]

Seymour Benzer [1921–2007]

Part 2: Through Sperry and Luria

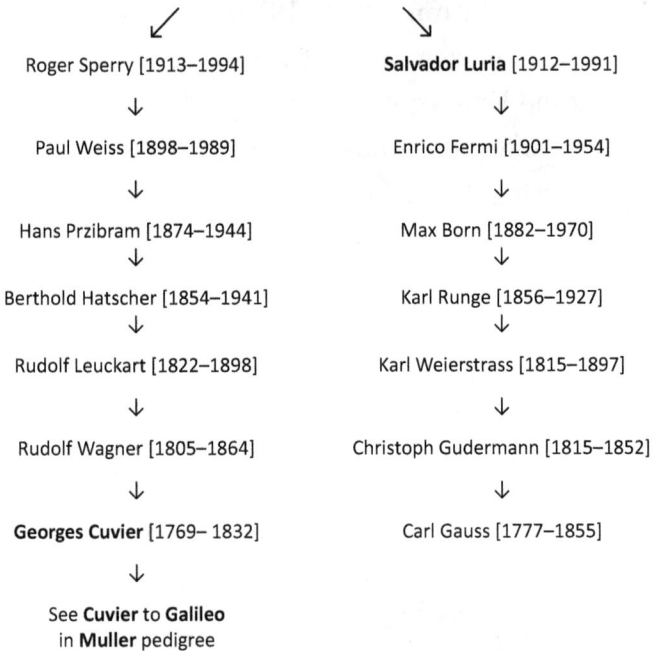

Roger Sperry [1913–1994]   **Salvador Luria** [1912–1991]

↓   ↓

Paul Weiss [1898–1989]   Enrico Fermi [1901–1954]

↓   ↓

Hans Przibram [1874–1944]   Max Born [1882–1970]

↓   ↓

Berthold Hatscher [1854–1941]   Karl Runge [1856–1927]

↓   ↓

Rudolf Leuckart [1822–1898]   Karl Weierstrass [1815–1897]

↓   ↓

Rudolf Wagner [1805–1864]   Christoph Gudermann [1815–1852]

↓   ↓

Georges Cuvier [1769– 1832]   Carl Gauss [1777–1855]

↓

See **Cuvier** to **Galileo**
in **Muller** pedigree

Legend for Intellectual Pedigree of Seymour Benzer

Seymour Benzer (1921–2007) was born in Brooklyn, New York, and died in Pasadena, California. He got his BA at Brooklyn College and then went to Purdue University for his PhD in solid state physics. His dissertation was on photoelectric effects on germanium. After he read **Schrödinger's** *What is Life?* Benzer decided to switch to the new molecular approach to genetics. He went to Caltech to study with **Max Delbrück**. Benzer studied the fine structure of genes using bacteriophage and showed genes that were divisible to hundreds of sites suggesting that he was mapping nucleotides in DNA. Later Benzer showed that fruit fly behavior could be studied at a genetic level by isolating induced mutants affecting behavior like mating or flying.

Karl Lark-Horovitz (1892–1958) was born in Vienna and died in West Lafayette, Indiana. His father was a dermatologist. He served in the signal corps in the Austrian army during WWI and was wounded in battle. He traveled to the United States and chose solid state physics for his research and was offered a job at Purdue University. He built a world class physics department there whose work led to the transistor. Two of his students went on to win Nobel Prizes, Edward Purcel and Julian Schwinger.

Franz Exner (1849–1926) was born and died in Vienna, Austria. His father was a professor who taught philosophy. Exner had two Nobel students — Lise Meitner and Erwin Schrödinger.

Victor von Lang (1838–1921) was born and died in Vienna. He was a founder of crystal studies in physics and wrote a monograph on optical properties and structure of crystals. He and Exner served as mentors to Erwin Schrödinger.

Max Delbrück (1906–1981) was born and raised in Berlin. He came from an academic family and on his mother's side he was a great-great-grandchild of Justus von Liebig. He studied atomic physics and got his PhD in 1930. He then shifted to biology after hearing lectures by **Niels Bohr** on photosynthesis arising from photon interactions in plastids and gene mutations arising in fruit flies from X-ray exposure. He learned genetics at the Kaiser Wilhelm Institute in Berlin with **Timoféev-Ressovsky**. He also met **H. J. Muller** there when Muller was on a Guggenheim Fellowship in 1932. Delbrück left Germany because he was opposed to Nazism. He moved to the United States and went to Caltech to explore using bacteriophage as model organisms for studying the gene at a molecular level. He was a founder of the Phage group and won a Nobel Prize for his contributions.

Niels Bohr (1885–1962) was a Danish physicist who was born and died in Copenhagen. His father was a physiologist and college professor. Bohr's work on the splitting of the atom provided the stimulus for making nuclear weapons. Bohr was Jewish and as the Nazis cracked down on Jews in Denmark, he escaped with thousands of other Jews by boats carrying them to Sweden. Bohr was mentored by **Christian Christenson.**

Milislav Demerec (1895–1966) was born in Kostajnike, Croatia, and died in Laurel Hollow, New York. He got his BA at Krizevci University in 1916 and emigrated to the United States in 1919. He got his PhD at Cornell in 1923 studying maize genetics with Emerson. He switched to fruit flies and he did research at Cold Spring Harbor Laboratory studying mutable genes. He founded Drosophila Information Service and became Director of the Laboratory after Blakeslee's retirement. He shifted to microbial genetics in the 1940s and used X-rays to produce strains of penicillin that greatly increased penicillin production. He showed genetic fine structure in bacteria while Benzer did so in bacteriophage.

Charles William Metz (1889–1975) was born in Wyoming and died at Woods Hole, Massachusetts. He got his BA at Pomona College in California and his PhD at Columbia University working with Morgan's fly lab. He studied *D. virilis* and obtained numerous mutations mapping them for comparison with *D. melanogaster.*

Max Delbrück (1906–1981) was born in Berlin, Germany, and died in Pasadena, California. His father was a historian and he was the great-great-grandson of **Justus von Liebig**. He became a physicist and got his PhD in 1930 at Göttingen. In 1932 he joined discussions on the gene at the Kaiser Wilhelm Institute with Timoféev-Ressovsky, Zimmer, and Muller. He left Germany in 1937 and came to the United States where he worked at Vanderbilt University and at Caltech. While at Caltech he did experiments with Ellis showing bacteriophage did not multiply exponentially, they emerged in single steps. He founded the phage group with Luria and Hershey, sharing a Nobel Prize for their work in establishing molecular genetics.

Rollins Emerson (1873–1947) was born in Pillar, New York, and died in Ithaca, New York. At age 7 his family moved to Nebraska. He liked agriculture and attended the University of Nebraska for a BA in 1897 and studied beans, confirming the Mendelian findings for many different traits. He then went to Harvard to study maize with E. M. East for his PhD. He joined Cornell's Botany Department and studied gene mutations in maize, making Cornell a world center for plant genetics.

Edward Murray East (1879–1947) was born in Du Quoin, Illinois, and died in Boston, Massachusetts. His father was a mechanical engineer. He worked on hybrid corn and showed dominance was the major reason for hybrid vigor after two inbred strains of corn were crossed. He applied this to the production of hybrid corn which vastly increased yields on farms. He also applied this to his views on eugenics. He felt inbreeding was bad if there were recessive lethal or detrimental genes present. But if there were no such genes, inbreeding could be used to produce strains of desired corn or other plants. He also applied this to the growing and later discredited eugenics movement, and his views were racist and elitist. Whites were superior and blacks were inferior. He also advocated eugenics favoring sterilization of social failures (paupers, psychotics, or mentally deficient).

Cyril George Hopkins (1866–1919) was born in Chatfield, California, and died in Gibraltar. He got his PhD at Cornell in 1898 studying agricultural chemistry. His dissertation was on "The chemistry of a corn kernel." He taught agricultural chemistry at the University of Illinois, showing how trace minerals and fertilizers could keep soil from being exhausted. In 1919 he wanted to help Greece recover from a war that was accompanied by crop failures. He was regarded as a hero for his contributions to that recovery, but he caught malaria and died in Gibraltar.

George Chapman Caldwell (1834–1907) was born in Framingham, Massachusetts, and died in Ithaca, New York. He received his BA at Harvard and then went to Germany to study with Robert Bunsen and Friedrich Wöhler. He returned with a PhD and during the Civil War served on the Sanitary Commission. He went to Cornell as its first appointed professor and built the chemistry department offering programs in both basic chemistry and agricultural chemistry.

Friedrich Wöhler (1800–1882) was born in Frankfurt, Germany, and died Göttingen, Germany. He studied with Berzelius and Gmelin. He obtained urea from ammonium cyanide and this was the first organic compound that was produced from inorganic components, establishing the existence of organic chemistry. Many chemists then believed organic compounds depended on a vital principle for their synthesis. Wöhler also discovered aluminum, titanium, yttrium, and beryllium as elements. He believed compounds could be broken down into their elements by appropriate chemical means.

Jöns Jacob Berzelius (1779–1848) was born in Östergötland, Sweden, and died in Stockholm, Sweden. His father was a school teacher but died when Jöns was one year old. His mother died when he was 7 years old. He was raised by relatives. He began his education at Uppsala University hoping to be a physician but switched to chemistry and physics. He showed that compounds could be separated by electric current into charged positive and negative ions. He determined the atomic weights of elements. He found silicon, selenium, thorium, and cerium. He also provided the nomenclature symbolism for elements and compounds 1 or 2 letters and a subscript for element in a compound (thus CO_2 is carbon dioxide; $CaCl_2$ is calcium dichloride).

Johan Afzelius (1753–1837) was born in Larv, Sweden, and died in Uppsala, Sweden. He got his PhD with Olof Bergman. He isolated formic acid from ants and showed it is the cause of its stinging bite. He isolated and named oxalic acid.

Bernhard Tollens (1841–1918) was born in Hamburg, Germany, and died in Göttingen, Germany. He studied with Karl Mobius as an undergraduate in 1857 and with Friedrich Wöhler in 1864 for his PhD. He spent his career working on carbohydrates, isolating and characterizing the sugars found in them.

Emil Erlenmeyer (1825–1909) was born in Taunusstein, Germany, and died in Aschaffenburg, Germany. His father was a Protestant minister. As an undergraduate he heard Liebig's lectures and decided he wanted to be a chemist. He studied with Bunsen and von Leibig and received his PhD in 1850. He was the first to use double and triple bonds in his chemical formulae for carbon atoms. He invented the Erlenmeyer flask. He analyzed alcohols and aldehydes and their composition.

Pedigree for Elizabeth Blackburn

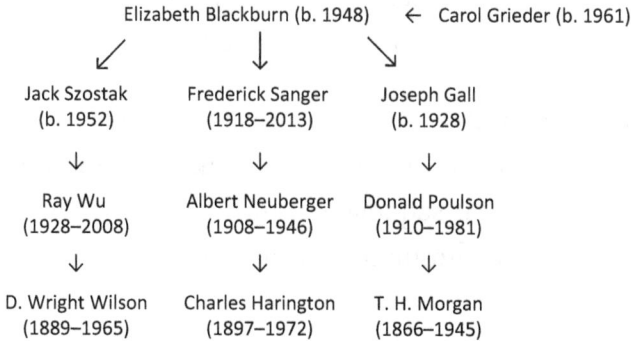

Elizabeth Blackburn (b. 1948)   ← Carol Grieder (b. 1961)

Jack Szostak (b. 1952)	Frederick Sanger (1918–2013)	Joseph Gall (b. 1928)
↓	↓	↓
Ray Wu (1928–2008)	Albert Neuberger (1908–1946)	Donald Poulson (1910–1981)
↓	↓	↓
D. Wright Wilson (1889–1965)	Charles Harington (1897–1972)	T. H. Morgan (1866–1945)

Legend for Pedigree of Elizabeth Blackburn

Elizabeth Blackburn (b. 1948) was born in Hobart, Australia. She received her PhD in 1975. She discovered the role of telomeres and telomerase structure, working with her student, Carol Grieder. Both Blackburn and Grieder with Jack Szostak shared a Nobel Prize for their work.

Carol Grieder (b. 1961) was born in San Diego, California. She worked with her mentor, Elizabeth Blackburn, on telomerase and the sequencing of telomere DNA. They worked out its genetic functions in cell division and cell aging.

Frederick Sanger (1918–2013) was born in Rendcomb and died in Cambridge, both in England. Sanger was a Quaker and became a chemist. He worked out the amino acid sequence and structure of insulin. For this he received his first Nobel Prize. He later worked out the techniques to determine the sequences of DNA molecules and for this received a second Nobel Prize in chemistry.

Albert Neuberger (1908–1996) was born in Hassfurt and died in London, both in England. He was a biochemist.

Charles Harington (1897–1972) was born in Llanerfyl, Wales, and died in London, England. He was a chemical pathologist. He worked out the structure of the hormone thyroxine.

Joseph Gall (b. 1928) was born Washington, D.C. He contributed to cell biology, showing DNA in a chromosome was a single molecule with a linear sequence of nucleotides. He analyzed the loops of lamp brush chromosomes and the coiling mechanisms of chromosomes.

Donald Poulson (1910–1989) was born in Idaho Falls, Idaho, and died in New Haven Connecticut. He was an embryologist who worked out the developmental events of fruit fly metamorphosis and the embryology of each stage.

Thomas Hunt Morgan [see pedigree]

Jack Szostack (b. 1952) was born in London and raised in Canada. His work on telomeres with Grieder and Blackburn earned him a Nobel Prize. He synthesized the first yeast artificial chromosome used in genetic engineering projects.

Ray Wu (1928–2008) was born in Beijing, China, and died in Ithaca, New York. He got his PhD in yeast genetics and has applied genetic engineering to agriculture.

D. Wright Wilson (1889–1965) was born in Knoxville, Iowa, and died in Philadelphia, Pennsylvania. He became a chemist and in WWI was burned by mustard gas while trying to test protective clothing. He taught at the University of Pennsylvania and worked on food chemistry using isotopes to study the metabolism of fats and carbohydrates.

Albert Blakeslee [1874–1954]
↓

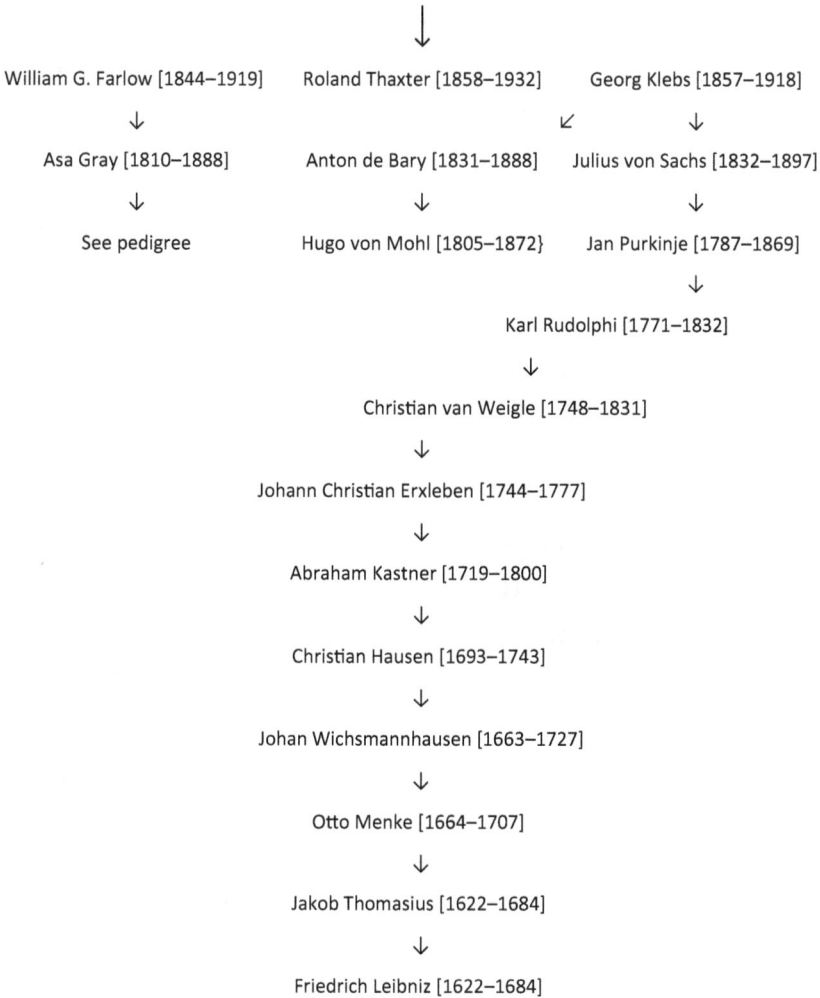

William G. Farlow [1844–1919]	Roland Thaxter [1858–1932]	Georg Klebs [1857–1918]
↓	↙	↓
Asa Gray [1810–1888]	Anton de Bary [1831–1888]	Julius von Sachs [1832–1897]
↓	↓	↓
See pedigree	Hugo von Mohl [1805–1872}	Jan Purkinje [1787–1869]

↓

Karl Rudolphi [1771–1832]

↓

Christian van Weigle [1748–1831]

↓

Johann Christian Erxleben [1744–1777]

↓

Abraham Kastner [1719–1800]

↓

Christian Hausen [1693–1743]

↓

Johan Wichsmannhausen [1663–1727]

↓

Otto Menke [1664–1707]

↓

Jakob Thomasius [1622–1684]

↓

Friedrich Leibniz [1622–1684]

Legend for Intellectual Pedigree of Albert Blakeslee

Albert Blakeslee (1874–1954) was born in Geneseo, New York, and died in Northampton, Massachusetts. He got his BA at Wesleyan and PhD at Harvard. He was the first to report sexual reproduction in the bread mold, Mucor. He spent a year in

Germany and shifted to plant aneuploidy and polyploidy. He used Jimson weed (Datura) to work out all the aneuploids for its n = 12 chromosomes. He also applied colchicine, with his student, O. Eigsti, to produce triploid offspring. He became director of Cold Spring Harbor Laboratories after Davenport retired.

William Gilson Farlow (1844–1919) was born in Boston, Massachusetts and died in Cambridge, Massachusetts. He got his MD at Harvard in 1870, worked with Asa Gray, and spent two years in Europe studying cryptogamic botany. He devoted his career equally to working out the relations of these plants and applying that knowledge to phytopathology.

Roland Thaxter (1858–1932) was born in Newtonville, Massachusetts. He was the third child. His father was a lawyer; and his mother was a poet. He started law school but switched to botany and took an interest in fungal infections of commercial vegetables — potatoes, onions, and lima beans. He also showed myxobacteria were colonial forms of bacteria and not fungi. He coined the term mycology for the study of fungi. He introduced the spray can with fungicides for protecting vulnerable plants from pathogens.

Julius von Sachs (1832–1897) was born in Breslau (now Wroclaw), Germany (now Poland), and died in Würzburg, Germany. His father was an engraver. When he was a young teen, his father died and two years later his mother and brother died of cholera. He was adopted by his father's friend, Jan Purkinje, who raised him in his home while von Sachs completed his college education. He got his PhD in 1856 and studied plant germination from seeds, plant nutrition, plant tropisms, and the history of botany. He showed that a leaf exposed to sun, bleached, and treated with iodine turned black from the starch it produced. A leaf kept in the dark, and bleached, did not turn black when dipped in iodine.

Jan Purkinje (1787–1869) was born in Libochoviche, Czechoslovakia, and died in Prague, Czechoslovakia. He got his MD at Charles University and studied the physiology of the senses. He described what are now called Purkinje cells in the cerebellum. He coined the terms plasma and protoplasm.

Milislav Demerec (1895–1966) was born in Kostajnike, Croatia, and died in Laurel Hollow, New York. He got his BA at Krizevci University in 1916 and emigrated to the United States in 1919. He got his PhD at Cornell in 1923 studying maize genetics with Emerson. He switched to fruit flies and he did research at Cold Spring Harbor Laboratory studying mutable genes. He founded Drosophila Information Service and became Director of the Laboratory after Blakeslee's retirement. He shifted to microbial genetics in the 1940s and used X-rays to produce strains of penicillin that greatly increased penicillin production. He showed genetic fine structure in bacteria while Benzer did so in bacteriophage.

Charles William Metz (1889–1975) was born in Wyoming and died at Woods Hole, Massachusetts. He got his BA at Pomona College in California and his PhD at Columbia University working with Morgan's fly lab. He studied *D. virilis* and obtained numerous mutations, mapping them for comparison with *D. melanogaster.*

Max Delbrück (1906–1981) was born in Berlin, Germany, and died in Pasadena, California. His father was a historian and he was the great-great-grandson of **Justus von Liebig**. He became a physicist and got his PhD in 1930 at Göttingen. In 1932 he joined discussions on the gene at the Kaiser Wilhelm Institute with Timofeef-Ressovsky, Zimmer, and Muller. He left Germany in 1937 and came to the United States where he worked at Vanderbilt University and at Caltech. While at Caltech he did experiments with Ellis showing bacteriophage did not multiply exponentially, they emerged in single steps. He founded the phage group with Luria and Hershey, sharing a Nobel Prize for their work in establishing molecular genetics.

Rollins Emerson (1873–1947) was born in Pillar, New York, and died in Ithaca, New York. At age 7 his family moved to Nebraska. He liked agriculture and attended the University of Nebraska for a BA in 1897 and studied beans, confirming the Mendelian findings for many different traits. He then went to Harvard to study maize with E. M. East for his PhD. He joined Cornell's Botany Department and studied gene mutations in maize, making Cornell a world center for plant genetics.

Edward Murray East (1879–1947) was born in Du Quoin, Illinois, and died in Boston, Massachusetts. His father was a mechanical engineer. He worked on hybrid corn and showed dominance was the major reason for hybrid vigor after two inbred strains of corn were crossed. He applied this to the production of hybrid corn which vastly increased yields on farms. He also applied this to his views on eugenics. He felt inbreeding was bad if there were recessive lethal or detrimental genes present. But if there were no such genes, inbreeding could be used to produce strains of desired corn or other plants. He also applied this to the growing and later discredited eugenics movement and his views were racist and elitist. Whites were superior and blacks were inferior. He also advocated eugenics favoring sterilization of social failures (paupers, psychotics, or mentally deficient).

Cyril George Hopkins (1866–1919) was born in Chatfield, California, and died in Gibraltar. He got his PhD at Cornell in 1898 studying agricultural chemistry. His dissertation was on "The chemistry of a corn kernel." He taught agricultural chemistry at the University of Illinois, showing how trace minerals and fertilizers could keep soil from being exhausted. In 1919 he wanted to help Greece recover from a war that was accompanied by crop failures. He was regarded as a hero for his contributions to that recovery, but he caught malaria and died in Gibraltar.

George Chapman Caldwell (1834–1907) was born in Framingham, Massachusetts, and died in Ithaca, New York. He received his BA at Harvard and then went to Germany to study with Robert Bunsen and Friedrich Wöhler. He returned with a PhD and during the Civil War served on the Sanitary Commission. He went to Cornell as its first appointed professor and built the chemistry department offering programs in both basic chemistry and agricultural chemistry.

Friedrich Wöhler (1800–1882) was born in Frankfurt, Germany, and died in Göttingen, Germany. He studied with Berzelius and Gmelin. He obtained urea from ammonium cyanide and this was the first organic compound that was produced from inorganic components, establishing the existence of organic chemistry. Many chemists then believed organic compounds depended on a vital principle for their synthesis. Wöhler also discovered aluminum, titanium, yttrium, and beryllium as elements. He believed compounds could be broken down into their elements by appropriate chemical means.

Jöns Jacob Berzelius (1779–1848) was born in Östergötland, Sweden, and died in Stockholm, Sweden. His father was a school teacher but died when Jöns was one year old. His mother died when he was 7 years old. He was raised by relatives. He began his education at Uppsala University hoping to be a physician but switched to chemistry and physics. He showed that compounds could be separated by electric current into charged positive and negative ions. He determined the atomic weights of elements. He found silicon, selenium, thorium, and cerium. He also provided the nomenclature symbolism for elements and compounds 1 or 2 letters and a subscript for element in a compound (thus CO_2 is carbon dioxide; $CaCl_2$ is calcium dichloride).

Johan Afzelius (1753–1837) was born in Larv, Sweden, and died in Uppsala, Sweden. He got his PhD with Olof Bergman. He isolated formic acid from ants and showed it is the cause of its stinging bite. He isolated and named oxalic acid.

Bernhard Tollens (1841–1918) was born in Hamburg, Germany, and died in Gottingen, Germany. He studied with Karl Mobius as an undergraduate in 1857 and with Friedrich Wöhler in 1864 for his PhD. He spent his career working on carbohydrates, isolating and characterizing the sugars found in them.

Emil Erlenmeyer (1825–1909) was born in Taunusstein, Germany, and died in Aschaffenburg, Germany. His father was a Protestant minister. As an undergraduate he heard Liebig's lectures and decided he wanted to be a chemist. He studied with Bunsen and von Liebig and received his PhD in 1850. He was the first to use double and triple bonds in his chemical formulae for carbon atoms. He invented the Erlenmeyer flask. He analyzed alcohols and aldehydes and their composition.

Theodor Boveri [1862–1915]

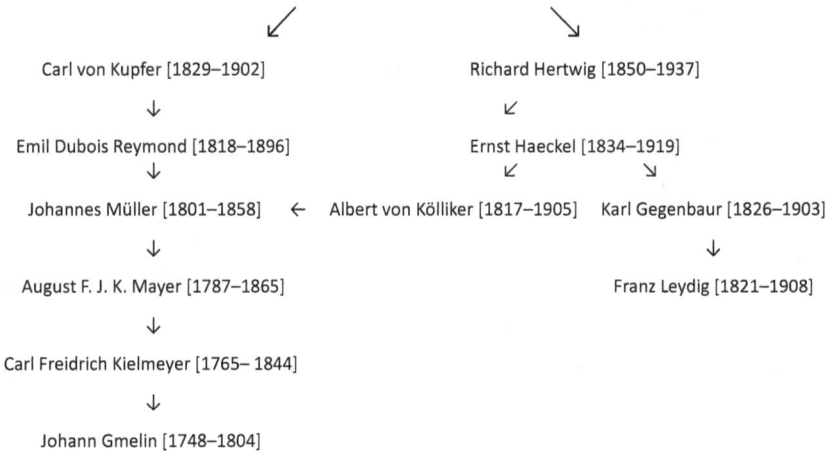

↙   ↘

Carl von Kupfer [1829–1902]   Richard Hertwig [1850–1937]

↓   ↙

Emil Dubois Reymond [1818–1896]   Ernst Haeckel [1834–1919]

↓   ↙   ↘

Johannes Müller [1801–1858]   ←   Albert von Kölliker [1817–1905]   Karl Gegenbaur [1826–1903]

↓   ↓

August F. J. K. Mayer [1787–1865]   Franz Leydig [1821–1908]

↓

Carl Freidrich Kielmeyer [1765– 1844]

↓

Johann Gmelin [1748–1804]

Legend for Pedigree of Theodor Boveri

Theodor Boveri (1862–1915) was born in Bamberg and died in Wurzburg, both in Germany. He obtained his MD in Munich and studied cell division by using experimental approaches. He agitated sea urchin fertilized eggs and caused abnormal chromosome numbers in them. The larvae either failed to develop or produced abnormal embryos. This led Boveri to propose the chromosome theory of heredity. His work both stimulated and supported the work of E. B. Wilson and his students in promoting that theory.

Carl von Kupffer (1829–1902) was born in Tukums, Latvia, and died in Munich, Germany. He was an anatomist specializing in the evolution of the head and spinal column. He also identified what are called Kupffer cells in the liver that are like white blood cells, scavenging bacteria and material bodies passing through the liver.

Emil Dubois Reymond (1818–1896) was born and died in Berlin, Germany. He studied the role of electricity in living organisms.

He provided the basis for the action potential in the passage of electricity from nerve to muscle. He also showed that muscles when exercised produced lactic acid.

Johannes Peter Müller (1801–1858) was born in Koblenz and died in Berlin, both in Germany. His father was a shoemaker. He started with an interest in becoming a priest but found the study of nature more exciting. He got an MD at Bonn and worked on the physiology of the senses. He also was distinguished as a morphologist, studying amphibian and reptilian similarities and differences. Many distinguished biologists were once his students.

August F. J. K. Mayer (1787–1865) was born in Gmund and died in Bonn, both in Germany. He studied the anatomy of primates and described the allantois as an embryonic membrane. He also embraced Goethe's naturphilosophie which saw life as spiritually directed and part of a universal oversoul.

Carl Freidrich Kielmeyer (1765–1844) was born in Bebenhausen and died in Stuttgart, both in Germany. He proposed a forerunner of Haeckel's recapitulation theory that present taxonomic groups reflect in their embryology the stages of evolution. He was also enamored by Goethe's naturphilosophie.

Johann Friedrich Gmelin (1748–1804) was born in Tübingen and died in Göttingen, both in Germany. He prepared the 13th edition of Linnaeus's *Systema Natura*. He studied irritability in plants identifying and classifying plants that trap insects, collapse leaves, or respond to light and gravity.

Richard Hertwig (1850–1937) was born in Friedberg and died in Icking, both in Germany. He was the first to describe zygote formation as a union of a single sperm with a single egg. He coined the term zygote. He also identified the coelom as an embryonic formation associated with the mesoderm of the early embryo.

Ernst Haeckel (1834–1919) was born in Potsdam and died in Jena, both in Germany. He coined the terms phylum, ecology, and phylogeny. He was a popularizer of science and proposed the phrase "ontogeny recapitulates phylogeny" or the past organization of life is reflected in the present embryology of living things.

Albert von Kölliker (1817–1905) was born in Zurich, Switzerland, and died in Würzburg, Germany. He provided many of the techniques for preparing tissues, embryos, and organs for microscopic analysis including dyeing, freezing, and sectioning specimens. He distinguished striated and smooth muscle and showed the former were syncytial and the latter were individual cells. He proved nerve fibers were outgrowths of neurons and not a secreted matrix of fibers. He also found and described mitochondria and noted they have their own membrane.

Norman Borlaug [1914–2009]
↓

Elvin Charles Stakman [1885–1975]

↓

Edward Monroe Freeman [1875–1954]

↓

Conway MacMillan [1867–1929]

↙   ↘

Ernst Athearn Bessey [1877–1957]   H. Marshall Ward [1854–1906]

↓   ↓

Georg Klebs [1857–1918]   Heinrich Anton de Bary [1831–1888]

↓   ↓

Julius von Sachs [1832–1897]   Hugo von Mohl [1805–1872]

Comments on Intellectual Pedigree of Norman Borlaug

Norman Ernest Borlaug (1914–2009) was born in Cresco, Iowa, and died in Dallas, Texas. He was of Norwegian ancestry and grew up in a Norwegian-American farming village in Iowa. He heard a lecture by Elvin Stakman when he was an undergraduate and it changed his life. It was on ways to increase farm yields by reducing insect and microbial infestation. He went to the University of Minnesota to study plant pathology and agriculture. He went to Mexico to try developing disease resistant, wind resistant, and high yield wheat. It took him about 20 years to succeed, and Mexico benefitted with surpluses. So did Pakistan and India. He is the father of the "Green Revolution" and won the Nobel Peace Prize for his efforts.

Elvin Charles Stakman (1885–1975) was born in Algoma, Wisconsin, and died in St. Paul, Minnesota. He worked out the life cycle of wheat rust and popularized the study of plant pathology.

Edward Monroe Freeman (1875–1954) was born and died in St. Paul, Minnesota. He founded the plant pathology department of the University of Minnesota.

Conway MacMillan (1867–1929) was born in Hillsdale, Michigan, and died in St. Paul, Minnesota. He worked out the alternation of generations in cryptogamic plant life in Minnesota.

Ernest Athearn Bessey (1877–1957) was born in Ames, Iowa, and died in East Lansing, Michigan. He was a mycologist and studied the mushrooms and fungal diseases of plant life. He went to Germany for his PhD.

Georg Klebs (1857–1918) was born in Nierenberg, Germany, and died (of the flu epidemic) in 1918 in Heidelberg, Germany. He studied the life cycles of cryptogamic plants, especially the early stages of development after spore or gamete production.

Julius von Sachs (1832–1897) was born in Breslow (then Prussia and now Poland) and died of diabetes in Würzburg, Germany. His parents died when he was starting college and his botany professor, **Purkinje**, adopted him. He was an outstanding botanist and plant physiologist. He proved that leaves made starch by exposing their chloroplasts to light. He worked out the transport of fluids in plants.

Harry Marshall Ward (1854–1906) was born in Hereford, England, and died in Babbecome on the English Channel. He was a plant pathologist and showed he could prevent coffee plants from infection by planting rows of leafy trees between fields of coffee plants. The trees prevented the fungal spores from being wind born to the unaffected plots.

Heinrich Anton de Bary (1831–1888) was born in Frankfurt, Germany, and died in Strasbourg. He worked out fungal life cycles, demonstrated sexuality among fungi, and coined the term symbiosis for the relation of algal and fungal components in lichens.

Hugo von Mohl (1805–1872) was born in Stuttgart, Germany, and died in Tübingen, Germany. His father was a statesman but von Mohl preferred the study of botany. He never married. He observed cell division in algae, coined the term protoplasm, and showed that the stomata of leaves opened and closed to regulate air intake.

Sydney Brenner [1927–2019]

↙ ↓ ↘

Cyril Norman Henshelwood [1897–1967]	Raymond Dart [1893–1968]	Fred Sanger [1918–2013]
↓		↓
Harold Brewer Hartley [1878–1972]		Albert Neuberger [1908–1996]
↓		
John Conroy [1845–1900]		
↓		
Augustus G. V. Harcourt [1834–1919]		
↓ ↘		
Benjamin Brodie [1817–1880]	William Esson [1838–1916]	
↓		
Justus von Liebig [1803–1873]		
↓		
Joseph Louis Gay-Lussac [1778–1850]		
↓		
Claude Louis Berthollet [1748–1822]		
↓		
Antoine Lavoisier [1742–1794]		
↓ ↘		
Nicolas Louis de Lacaille [1713–1762]	Etienne B. de Condillac [1714–1780]	
↓	↓	
Jacques Cassini [1677–1756]	John Locke [1632–1704]	

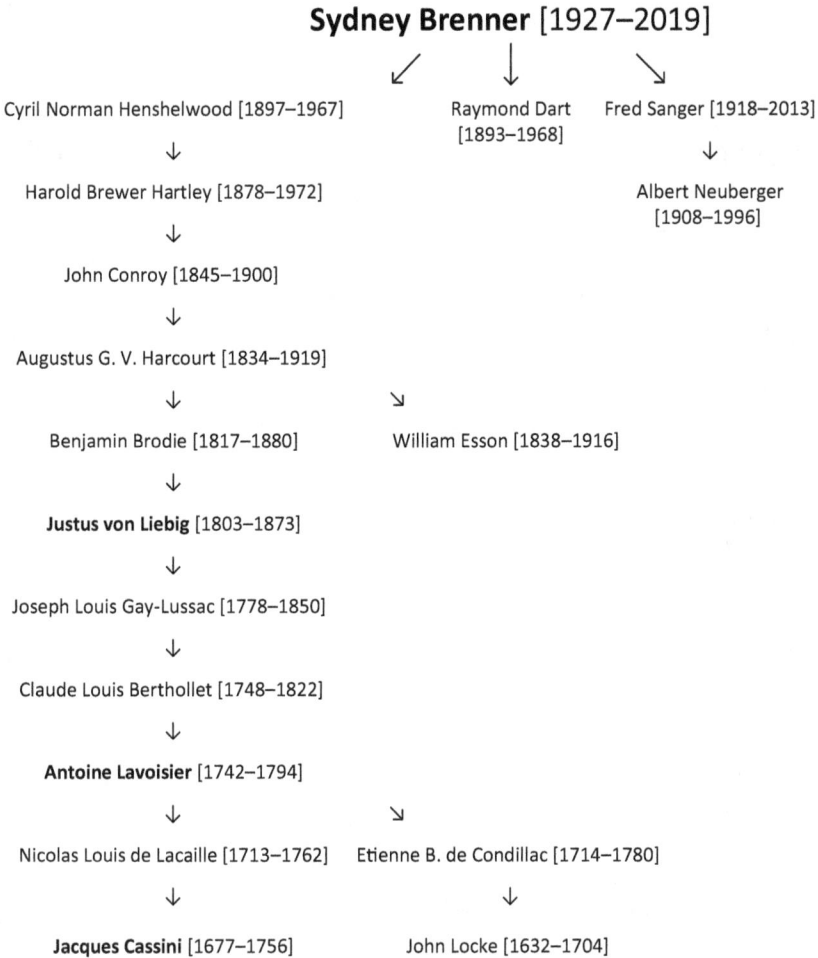

Legend for Intellectual Pedigree of Sydney Brenner

Sydney Brenner was born 1927 in Germiston, South Africa, and he grew up as a child of immigrant parents from the Baltic States. He died in 2019 in Singapore. His father was a cobbler. He attended the University of Witwatersrand and received a BS and MS in anatomy and physiology. Paleontologist **Raymond Dart** was his mentor for anatomy. He shifted to cytogenetics for his second MS degree and won a fellowship to attend Exeter College in Oxford where he received a PhD under **Cyril Norman Hinshelwood**. His dissertation was on the chemical kinetics of phage infection and resistance in *E. coli*. Brenner shifted to the Medical Research Council and worked with **Francis Crick** on DNA and its relation to coding. He was one of the first to propose the existence of messenger RNA. He showed that the genetic code could be worked out from mutations in DNA sequences leading to precise substitutions of amino acids in the proteins specified by those genes. His discovery of frame shift mutations provided insights into the functions of genes at the molecular level. Later Brenner chose *Caenorhabditis elegans* as a model organism for the molecular biology of development in eukaryotic organisms. He was awarded a Nobel Prize for his work on the molecular basis of development in these roundworms. He moved to the Salk Institute in La Jolla, California and retired there.

Raymond Dart (1893–1968) was born in Brisbane, Australia, and died in Johannesburg, South Africa. His father was a farmer. He attended Queensland College for his BSc and attended medical school. In WWI he was a medic. He completed his MD after the war and moved to South Africa. There he discovered and described *Australopithecus africanus* and argued that Africa was the source of the human genus. Australopithecus had too large a brain size to be considered an ape.

Frederick Sanger (1918–2013) was born in Gloucestershire, England, and died in Cambridge, England. His father was a physician and a Quaker. Sanger was a pacifist and both his parents died of cancer while he was in college. He went on to win two Nobel Prizes, the first for sequencing the amino acids in insulin and the second for working out a method to sequence nucleotide pairs in DNA.

Albert Neuberger (1908–1996) was born in Hassfurt, Germany, and died in Hampstead, England. He was Jewish and fled Germany in 1933 when Hitler came to power. He settled in London where he obtained his PhD studying glycoproteins and the metabolism of nitrogen and other components of potatoes.

Cyril Norman Hinshelwood (1897–1967) was born and died in London. He studied physical chemistry and applied his findings to the living cell. His PhD was from Oxford. His contributions to chemical kinetics led to his earning a Nobel Prize in chemistry.

Harold Brewer Hartley (1878–1972) was born in London. He taught and studied physical chemistry at Oxford. He was a Brigadier General in WWI and worked on chemical warfare.

John Conroy (1845–1900) was born in London and died in Rome, Italy. His father was a baronet and he went to Eton and Oxford for his education. He liked the natural sciences and became a chemist, working on optical measurements of molecular reactions.

Elof Axel Carlson [b. 1931]
↙   ↘

Morris Gabriel Cohen [1900–1973]

Hermann J. Muller [1890–1967]
↙   ↓

Thomas H. Morgan [1866–1945]
↙   ↓

Edmund B. Wilson [1856–1939]
↙

H. Newell Martin [1848–1896]	William K. Brooks [1848–1908]
↓	↓
Michael Foster [1836–1907]	Louis Agassiz [1807–1873]
↓	↓
Thomas H. Huxley [1825–1895]	Georges Cuvier [1769–1832]
↓	↓
Charles Darwin [1809–1882]	Ignaz Döllinger [1770–1841]
↓	↓
Adam Sedgwick [1785–1873]	Antonio Scarpa [1752–1832]
↓	↓
Thomas Jones [1756–1807]	Giovanni Morgagni [1682–1771]
↓	↓
Thomas Postlethwaite [1731–1798]	Antonio Valsalva [1666–1723]
↓	↓
Stephen Whisson [1710–1783]	Marcello Malpighi [1628–1694]
↓	↓
Walter Taylor [1700–1743]	Giovanni Borelli [1608–1679]
↓	↓
Robert Smith [1689–1768]	Benedetto Castelli [1578–1643]
↓	↓
Roger Cotes [1682–1716]	Galileo Galilei [1564–1642]
↓	↓
Isaac Newton [1642–1726]	Ostilio Ricci [1540–1603]
	↓
	Niccolò Tartaglia [1499–1557]

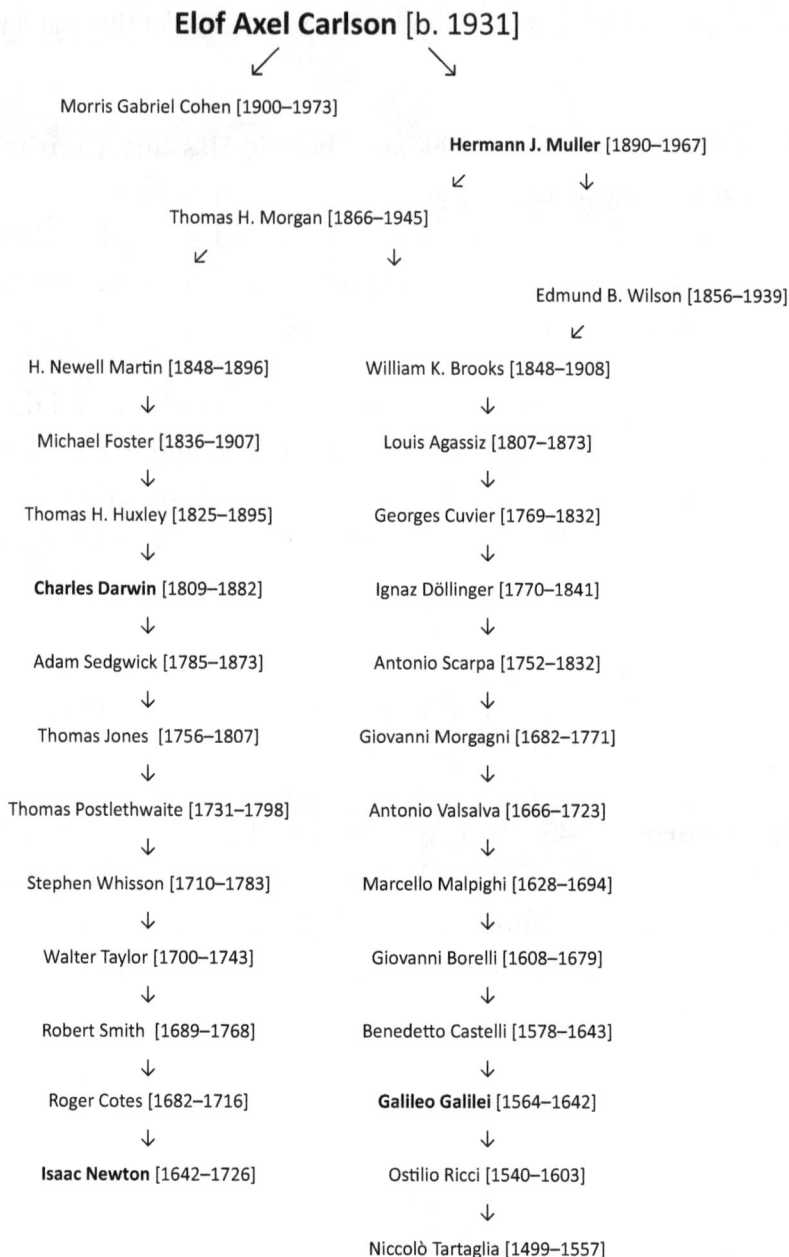

Legend for Intellectual Pedigree of Elof Carlson

See Topics 4–8 for the commentaries on each of Carlson's mentoring predecessors.

William Ernest Castle [1867–1962]

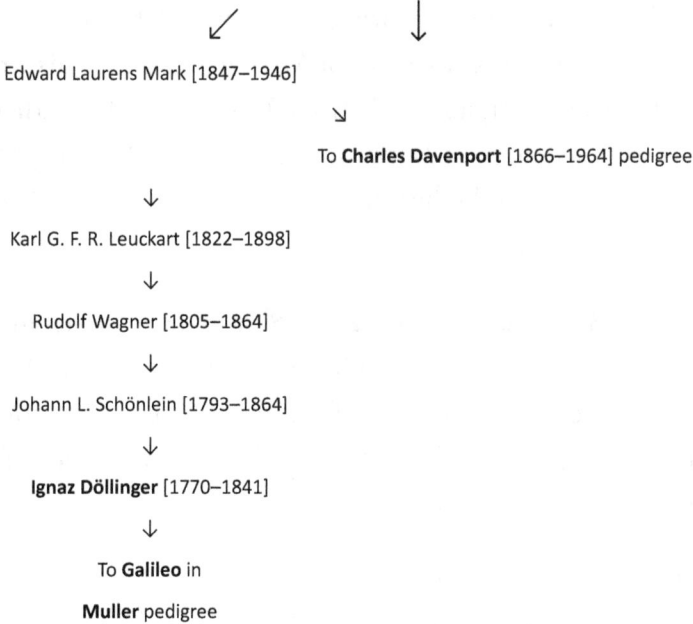

↙ ↓

Edward Laurens Mark [1847–1946]

↘

To **Charles Davenport** [1866–1964] pedigree

↓

Karl G. F. R. Leuckart [1822–1898]

↓

Rudolf Wagner [1805–1864]

↓

Johann L. Schönlein [1793–1864]

↓

Ignaz Döllinger [1770–1841]

↓

To **Galileo** in

Muller pedigree

Legend for Intellectual Pedigree of William Castle

William Ernest Castle (1867–1962) was born in Ohio and died at Berkeley, California. He went to Denison College and majored in classics. Later he switched to biology first in embryology and then in genetics after the Mendelian rediscovery in 1900. He used mammals especially mice and guineas pigs. He taught at Harvard. He used fruit flies to study minor traits (like wing veins and wing length) in fruit flies in 1905. He believed variable traits had variable genes, a view challenged by the work of Morgan and his students, especially Muller, who favored a model of chief genes and modifiers for the variation. Eventually Castle did a decisive experiment in hooded rats and confirmed the work of the fly lab.

Edward Laurens Mark (1847–1946) was born in Hamlet, New York, and died in Cambridge, Massachusetts. He received his BA in 1871 at the University of Michigan and got his PhD in 1876 in Leipzig, Germany with **Karl Leuckart**. He studied the life cycle and histology of *Limax* a slug (terrestrial gastropod). As a professor at Harvard he had among many students, **Davenport**, **Castle**, and **Jennings**.

Karl G. F. R. Leuckart (1822–1898) was born in Helmstedt, Germany, and died in Leipzig, Germany. He was a parasitologist working out life cycles and hosts for human tapeworms, trichinosis, and for sheep liver flukes. His work led to inspections of meat for parasites in industrialized countries.

Rudolf Wagner (1805–1864) was born in Bayreuth, Germany, and died in Göttingen, Germany. He described the germinal vesicle of the ovum in mammalian ovaries. It was later shown to be an arrested stage in prophase I of meiosis and it resumes cell division (when a surge of luteinizing hormone occurs during mature egg formation). He was a vitalist who followed Goethe's view of biology.

Johann L. Schönlein (1793–1864) was born and died in Bamberg, Germany. He broke tradition and lectured in German instead of Latin. He identified the cause of ringworm as a scalp fungal parasite *Trichophyton schönlandii*). He also gave the name tuberculosis to what was then called consumption.

Luigi Cavalli–Sforza [1922–2018]

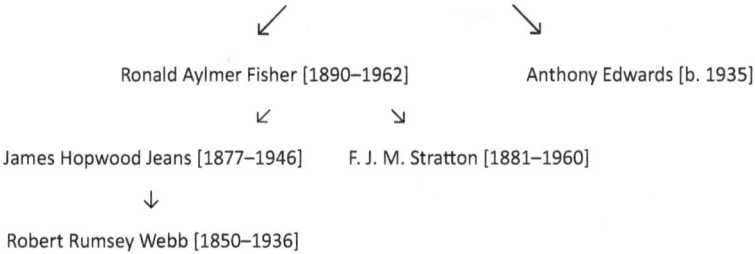

↙   ↘

Ronald Aylmer Fisher [1890–1962]   Anthony Edwards [b. 1935]

↙   ↘

James Hopwood Jeans [1877–1946]   F. J. M. Stratton [1881–1960]

↓

Robert Rumsey Webb [1850–1936]

Legend for Pedigree of Luigi Cavalli-Sforza

Luigi Luca Cavalli-Sforza (1922–2018) was born in Genoa, Italy, and died in Belluno, Italy. His career focused on population genetics. He applied it to human populations, first with blood groups and then with DNA analysis of populations whose migrations he traced. Europeans have 2/3 Asian and 1/3 African ancestry. He also proposed co-evolution of both ethnic differences and cultural differences. He taught in Parma, Pavia, and Stanford University.

Ronald Aylmar Fisher (1890–1962) was born in London, England, and died in Adelaide, Australia. His father was an auctioneer of fine art. Despite failing eyesight, he became a founder of population genetics and made major contributions to statistics and to the New Synthesis that merged classical genetics, cytology, paleontology, and population genetics with Darwinian natural selection. Fisher favored eugenics and conservative politics. He was also interested in parapsychology.

Anthony Edwards (b. 1935) was born in London, England. His father was a surgeon and his older brother John Edwards, was a medical geneticist known for first describing Edwards' syndrome (trisomy 18). Anthony Edwards studied phylogenetics and applied molecular sequencing of DNA and proteins to taxonomy and evolution. His most famous studies have traced human migrations out of Africa and across the continents.

James Hopwood Jeans (1877–1946) was born in Southport, England, and died in Dorking, England. His father was a Parliamentary reporter and author. James Jeans was gifted in mathematics and applied his knowledge to astronomy. He proposed and supported the "steady state" model of the universe in which the universe is undying and constantly recycles its energy and matter. The Big Bang model provided more testable evidence and is presently favored. Jeans favored transcendental views of the universe and claimed that the universe is closer to a great thought than a great machine.

Frederick John Marrian Stratton (1881–1960) was born near Birmingham, England, and died in Cambridge, England. He was an astronomer, applying his skills to solar physics. In World War I he served in the signal corps. He took an interest in parapsychology research.

Robert Rumsey Webb (1850–1936) was born in Monmouth, Wales, and died in Cambridge. He chose to be a mathematics coach (wrangler) for the mathematics "tripos" in Cambridge.

Ralph Cleland [1892–1971]

↓

Bradley M. Davis [1871–1957]

↓

Otto Renner [1883–1960]

↙   ↓   ↘

Karl von Goebel [1855–1932]   Erwin Baur [1875–1933]   Wilhelm Pfeffer [1845–1920]

↓   ↓

Anton de Bary [1831–1888]   Nathaniel Pringsheim [1823–1894]

↓

Hugo von Mohl [1805–1872]

Legend for Ralph Cleland Intellectual Pedigree

Ralph Cleland (1892–1971) was born in LeClair, Iowa, and died in Bloomington, Indiana. His father was a Presbyterian minister who moved to Philadelphia when young Cleland was still an infant. Cleland attended the University of Pennsylvania majoring in classics and minoring in history with a few botany courses added because he loved nature walks. He graduated Phi Beta Kappa and was drafted in World War I going to France when he was struck with the flu epidemic. He was discharged and got his PhD in botany with **Bradley Davis** who set him to work on the genetics of Oenothera, the evening primrose. Cleland discovered the formation of rings of translocated chromosomes which integrated the peculiar modes of inheritance of this genus of plants. It included balanced lethal mutations, pure breeding hybrids, and an excessively high mutation rate. Most of Cleland's career was at Indiana University in Bloomington.

Bradley Moore Davis (1871–1957) was born in Chicago, Illinois, and died in Portland, Oregon. He got his BA at Stanford University and then went to Harvard for his MA and PhD. He also visited scholars in Bonn, Germany and Naples, Italy. His PhD was on red algae but his main contributions to botany were on the genetics of Oenothera and its species.

Otto Renner (1883–1960) was born in Neu-Ulm, Germany, and died in Munich, Germany. He studied with **Goebel** in Munich and **Pfeffer** in Leipzig. Most of his work was on Oenothera. He identified complexes transmitted as units. These later were shown by Cleland to be ring-shaped groups of translocations. In 1945 his laboratory was destroyed by Allied bombs and nine of his staff were killed.

Karl von Goebel (1855–1932) was born in Baden, Germany, and died in Munich, Germany. He was originally seeking an education in theology to become a minister, but lectures of **Wilhelm Hofmeister** turned him on to botany. He got his PhD with **Anton de Bary** on plant sexuality and development and became immersed in comparative morphology and comparative physiology of plants from fungi to angiosperms. He died of heart failure shortly after breaking his shoulder in a fall while collecting specimens in the Alps.

Anton de Bary (1831–1888) was born in Frankfurt, Germany, and died in Strasbourg, France. His father was an MD. He got his MD in 1853 in Berlin with a thesis on plant sexuality. He studied slime molds, wheat infections by rust (Puccinia), potato blight (Phytophthora), and showed plant diseases came from fungi with a life cycle and they were not products of decaying plant cells.

He identified lichens and gave the name symbiosis for their dual natures of having algal and fungal cells.

Hugo von Mohl (1805–1872) was born in Stuttgart, Germany, and died in Tübingen, Germany. His father was a statesman in Würtemberg. Von Mohl never married. He learned most of his botany on his own. In 1835 he observed cell division but did not see the chromosomes or stages of cell division. He rejected the Schleiden and Schwann theory of free formation of cells and argued the cell never dissolved during division. He worked out the opening and closing of stomata in leaves. He also showed how cells form the tissues of plant vessels and fibrous structures.

Wilhelm Pfeffer (1845–1920) was born in Grebenstein, Germany, and died in Leipzig, Germany. He studied chemistry and pharmacy in Göttingen with **Wöhler** and learned botany from **Pringsheim**. He taught at Basel, Tübingen, and Leipzig. His research was on plant physiology. He devised instruments to measure osmosis in plant cells, and plant movement. He was the first to use time lapse photography to show plant growth.

Carl Correns [1864–1933]

↓

Karl Nägeli [1817–1891]

↓

Alphonse de Candolle [1806–1893]

↓

Augustin P. de Candolle (1778–1841)

↓

Jean P. Vaucher [1763–1841]

Legend for Carl Correns Intellectual Pedigree

Carl Correns (1864–1933) was born in Munich, Germany, and died in Berlin, Germany. He was an orphan and raised by his aunt. He studied botany at Tübingen. In 1900 he published his confirmation of Mendel's laws of heredity. He shifted to working with *Mirabolis jalapa*, the 4 o'clock flowering plant. He found that the chloroplasts followed a maternal (ovule cytoplasmic) inheritance. Pollen did not carry the chloroplasts to the ovules they fertilized. In 1913 Correns was named Director of the Kaiser Wilhelm Institute in Berlin-Dahlem.

Karl Nägeli (1817–1891) was born in Kilchberg, Switzerland, and died in Munich, Germany. He became a botanist and identified the hereditary material of plants as idioplasm. He introduced the terms meristem, phloem, and xylem when describing plant anatomy and the relation of these structures to their functions in plant physiology and development. He also corresponded with **Mendel** and exchanged seeds with him to study each other's work on transmission of traits.

Alphonse de Candolle (1806–1893) was born in Paris, France, and died in Geneva, Switzerland. He started out in law and switched to botany, his father's profession. He is a founder of geographic botany (1855) and his writings influenced **Nikolai Vavilov** in his studies of the centers of origins of domesticated plants.

Augustin P. de Candolle (1778–1841) was born and died in Geneva, Switzerland. He was first to describe a biological clock in plants by studying their motions with and without sunlight. He introduced the idea of homologous structures (e.g., flowering plant petals and leaves). He described "nature's war" as a competition for space, community, and food but did not apply it to ideas of evolution as Darwin did. He also shifted plant classification from arbitrary features (e.g., petal or stamen number) to structural features based on plant anatomy.

Jean P. Vaucher (1763–1841) was born and died in Geneva, Switzerland. His father was a carpenter. He was a Protestant pastor and taught church history and botany. His specialty was algae. He was the first to suggest conjugation (e.g., in spirogyra) was a sexual process leading to spore production.

F. A. E. Crew [1886–1973]

↙   ↘

Arthur Darbyshire   Edward Sharpey-Schafer [1850–1935]
[1879–1915]
↓   ↓

Raphael Weldon   William Sharpey [1802–1880]
[1860–1906]
↓   ↙   ↘

Francis Balfour   John Barclay   Guillaume Dupuytren [1777–1835]
[1851–1882]   [1758–1826]

↓   ↓

Michael Foster   Andrew Marshall
[1836–1907]   [1779–1858]
↓

See intellectual
pedigree of **H. J. Muller**
Foster to **Newton**

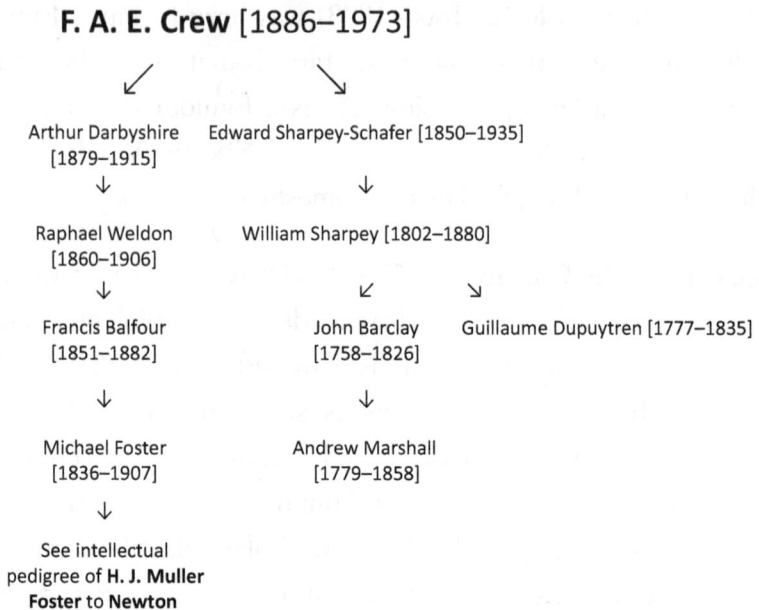

Legend for Intellectual Pedigree of F. A. E. Crew

Francis Albert Eley Crew (1886–1973) was born in Tipton and died in Sussex, both in England. His father was a grocer. Crew studied medicine at Edinburgh and in WWI he served in the medical corps. He got his DSc in developmental genetics, studying the embryology of the reproductive system in vertebrates. He did research on intersexual conditions and their physiological and anatomic features. He headed the animal genetics unit of the University of Edinburgh and during WWII he hosted **H. J. Muller**, **Charlotte Auerbach**, and **Guido Pontecorvo**, all victims of Fascist or Communist attacks on genetics.

Arthur Dukinfield Darbishire (1879–1915) was born in Kensington and died in London, both in England. His father was an MD. Darbishire was a student of embryologist **Francis Balfour** at Oxford. He adopted the biometric approach to evolution that

Balfour favored but after the rediscovery of Mendelism in 1900 he began breeding mice, poultry, and other small vertebrates. He argued that both Mendelism and biometrics were valid for evolutionary studies in his book, *Breeding and the Mendelian Discovery*. In 1915 he enlisted in the artillery but contracted meningitis and died in London.

Walter Frank Raphael Weldon (1860–1906) was born in London and died in Oxford, both in England. He was a founder of the field of biometrics with **Karl Pearson** and **Francis Galton**. He also did research on embryology and its ties to evolution through phylogeny. The "biometric school" was opposed by **William Bateson** who favored discontinuous evolution through sudden mutations that were meristic (duplication of organs) or homeotic (displacement of organs in the body) mutations.

Francis Maitland Balfour (1851–1882) was born in Edinburgh, Scotland, and died in Chamonix, France. He studied embryology, especially the urogenital system in vertebrate embryos. He died young, falling to death while attempting to climb Mont Blanc in the Alps.

Michael Foster (1836–1907) was born in Huntingdon, England, and died in London, England. He was a student of **Thomas Henry Huxley** and studied physiology and its relation to evolution.

Edward Sharpey-Schafer (1850–1935) was born in London and died in North Berwick, both in England. His father was German, and his mother was English. He got his MD at University College in London. He studied with **William Sharpey** and added Sharpey's name to his own after Schafer's son was killed in WWI. Sharpey-Schafer improved histological techniques for the study of tissues and cell organelles. He wrote a text in histology and he is also known for publishing a paper on how to provide artificial respiration to victims of choking or sudden death.

William Sharpey (1802–1880) was born in Arboath, Scotland and died in London. His father was a shipowner. He studied surgery in Edinburgh and then studied in Padua, Berlin, and Heidelberg and got his MD in Paris with Dupuytren. Sharpey's fibers in anatomy are associated with the attachment of tendons to bone.

Andrew Marshall (1779–1858) was an anatomist who began his medical career as a ship's mate and spent most of his career in the Royal Navy. He got his MD at Glasgow and founded the Belfast Medical Society.

Guillaume Dupuytren (1777–1835) was born in Pierre-Buffière and died in Paris, both in France. He was one of the most successful surgeons who became wealthy from his skills. He is best known for Dupuytren's contracture of the fingers and for treating Napoleon's hemorrhoids (he also wrote a treatise on constructing an artificial anus). He was a difficult critic and known as "first among surgeons and least among men."

John Barclay (1758–1826) was born in Perthshire and died in Edinburgh, both in Scotland. His father was a farmer. He became a minister and studied natural history and became a tutor for the children of wealthy patrons.

FRANCIS CRICK [1916–2014]

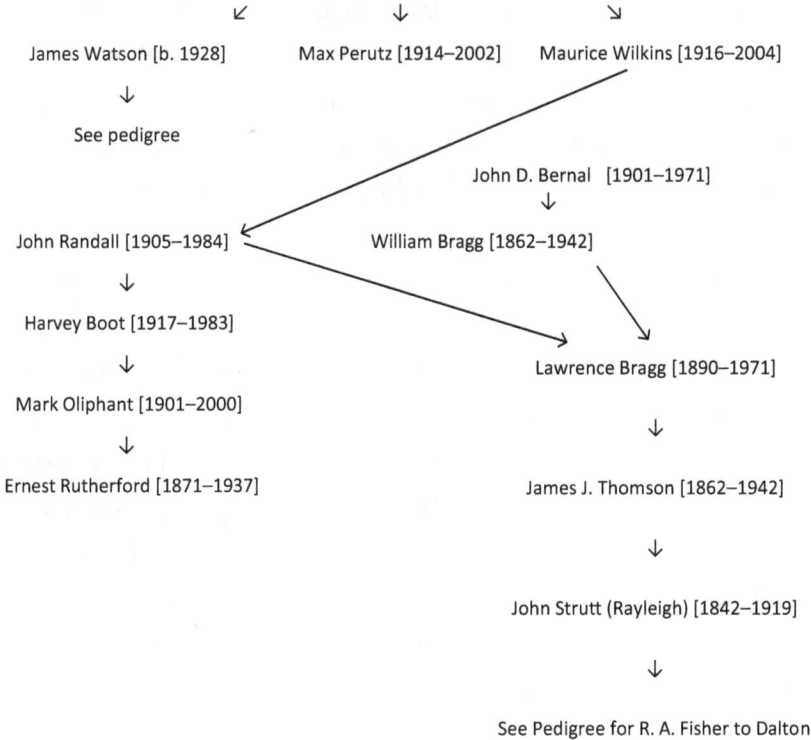

↙   ↓   ↘

James Watson [b. 1928]   Max Perutz [1914–2002]   Maurice Wilkins [1916–2004]

↓

See pedigree

John D. Bernal [1901–1971]

↓

John Randall [1905–1984]   William Bragg [1862–1942]

↓

Harvey Boot [1917–1983]

↓   Lawrence Bragg [1890–1971]

Mark Oliphant [1901–2000]

↓   ↓

Ernest Rutherford [1871–1937]   James J. Thomson [1862–1942]

↓

John Strutt (Rayleigh) [1842–1919]

↓

See Pedigree for R. A. Fisher to Dalton

Legend for Pedigree of Francis Crick

Francis Harry Compton Crick (1916–2004) was born in Weston Favell, England, and died in San Diego, California. His father was a shoe manufacturer. He became an atheist at age 12. He attended the University of London for a BSc in physics and during WWII worked on the detection and placement of mines. He switched to genetics after reading Schrödinger's *What is Life?* He got his PhD with **Perutz** for crystallographic studies of proteins. His work on DNA with **Watson** took place before he completed his PhD. He moved to the Salk Institute and devoted his later career to the neuro-biological study of consciousness.

James Watson (b. 1928) was born in Chicago, Illinois. His father was a tax collector. He was precocious and a Quiz Kid partisan on national radio. He entered the University of Chicago at age 15. He intended ornithology as a career but after reading Schrödinger's *What is Life?* he shifted to genetics and went to Indiana University for his PhD. He enjoyed **Muller's** courses and conception of the gene but did his dissertation research on radiation damage to viruses with **Luria** as his mentor. He learned of DNA crystallography while on a postdoctoral stay in Europe and switched to Cambridge and there met **Crick**. Together they solved the structure of DNA and applied it to gene replication, information storage, and mutation. Watson returned to the United States and did research at Harvard before becoming Director of Cold Spring Harbor Laboratories. He has written about 15 books and participated in organizing the human genome project. He remains controversial for his outspoken self-assessment and beliefs about genetics in human affairs, including gender, race, and abilities.

John Randall (1905–1984) was born in Newton-le-Willows, England, and died in Edinburgh, Scotland. His father was a nurseryman. He studied physics with Lawrence Bragg and during WWII he worked with **Boot** and **Oliphant** on the magnetron, a device that they applied to radar and to microwave instruments such as ovens.

Harvey Boot (1917–1983) was born in Birmingham and died in Cambridge, both in England. He got his PhD in physics with **Oliphant** working on radar.

Mark Oliphant (1901–2000) was born in Kent town, Australia, and died in Canberra, Australia. He discovered tritium, a form of hydrogen with one proton and two neutrons. He showed that the fusion of deuterium (one proton and one neutron) with hydrogen (one proton) releases energy. His PhD work was done with **Rutherford**. During WWII his work was used as the basis for making the hydrogen bomb.

Ernest Rutherford (1871–1937) was born in Brightwater, New Zealand, and he died in Cambridge, England. He founded the field of nuclear physics. He introduced the half-life measurement of radioactive decay. He discovered the atom contains particles, including one or more nuclear protons and that alpha rays from radioactive decay was composed of helium atoms. He worked on Sonar during WWI for detecting submarines. He also showed that atoms may contain one or more neutrons. This generated the model of a nuclear atom with surrounding electrons.

Max Perutz (1914–2002) was born in Vienna, Austria, and died in Cambridge, England. His parents were Jewish and converted the family to Catholicism. Later, Max became an atheist but never attacked religion (he felt we should act as if there was a God). His parents escaped in 1938 to Switzerland and Perutz learned the physics of water, snow, and ice, and studied glaciology. In WWII, while in England studying physics with **Bernal**, he applied this knowledge of the physics of water to problems like surviving in ice caverns. After the war he shifted his crystallography to the study of hemoglobin and worked out its structure.

John Desmond Bernal (1901–1971) was born in Nenagh, Ireland, and died in London, England. He was of mixed Irish, Sephardic Jewish, and Hispanic Catholic heritage on his paternal side. His mother was American. He got his PhD with **William Bragg** at the Faraday laboratory in London. He applied crystallography to graphite, steroids, cholesterol, and tobacco mosaic virus.

His friends called him "Sage". His communist political outlook limited his career and his contributions during the war effort in WWII.

William Bragg (1862–1942) was born in Wigton, England, and died in London, England. His father was a merchant marine officer but when William's mother died when he was 7 years old, he was raised by his uncle's family. He went to Cambridge for his physics and mathematics training and was mentored by **J. J. Thomson**. He moved to Adelaide in Australia and worked on radio communication and returned to study X-rays. He devised the instrument to produce X-ray diffraction pictures of crystals (his son, Lawrence, being a co-author) for which they shared a Nobel Prize. Another son was killed in WWI two months before the Nobel award was announced.

Maurice Wilkins (1916–2004) was born in Pongaroa, New Zealand, and died in London, England. His father was an MD. His family moved to Birmingham, England when he was 6 years old. He received a Cambridge BA and a PhD with **Randall** as his mentor in 1940. He worked on the Manhattan Project during the war. At King's College in London he worked on X-ray diffraction of DNA. **Rosalind Franklin** was added by Randall who did not tell Wilkins she would also be working on DNA. While Wilkins was first to use DNA and produce a picture, Franklin's was sharper and proved (to Crick who examined the photograph) that the molecule was a double helix. Wilkins, Crick and Watson shared a Nobel for the DNA structure. The Nobel Rule of 4 (not permissible) was not invoked, however, because Franklin had died of cancer before the award was made. No Nobel is posthumously awarded.

Lawrence Bragg (1890–1971) was born in Adelaide, Australia, and died in Waldringfield, England. He was the son of **William**

Bragg. He studied physics at the University of Adelaide and after he returned to England, he studied physics at Cambridge. He conceived of X-ray crystallography photography and its mathematical interpretation of crystal structure and his father built the machine to do this. For this work they shared a Nobel Prize. During WWI he devised a heat detection system to reveal explosions at a distance whose sound could not be amplified.

Joseph John Thomson (1856–1940) was born in Manchester, England, and died in Cambridge, England. His father owned an antique bookstore. He was an Anglican throughout his life. His father died when he was 14 and Thomson went to Cambridge to study mathematics and physics. In 1883 he received his MA and worked on the passage of electricity through gases which led him to the discovery of an atomic particle that he called a corpuscula and later renamed electron. He showed the electron was 1/1000th the size of an atom and hydrogen atoms had only one electron.

John Strutt (Baron Rayleigh III) (1842–1919) was born in Maldon, England, and died in Witham, England. He received his BA at Cambridge and studied acoustics. Among his findings was the answer to why the sky is blue. It is caused by small particles in the atmosphere that absorb the red light from the sun. He also studied the density of gases and showed that atmospheric nitrogen was heavier than predicted from isolation by chemical reactions. He traced this to the presence of another gas, argon. For this discovery he shared a Nobel Prize.

James F. Crow [1916–2012]

John Thomas Patterson [1878–1960]　　　Hermann J. Muller [1890–1967]

↓　　　　　　　　　　　　　　　　　↓

Charles Otis Whitman [1842–1910]　　　See **Muller** pedigree to **Newton**

↓

Karl Leuckart [1822–1898]

↓

Rudolf Wagner [1805–1864]

↓

Johann Schönlein [1793–1864]

↓

See **Muller** pedigree
from **Döllinger** to **Galileo**

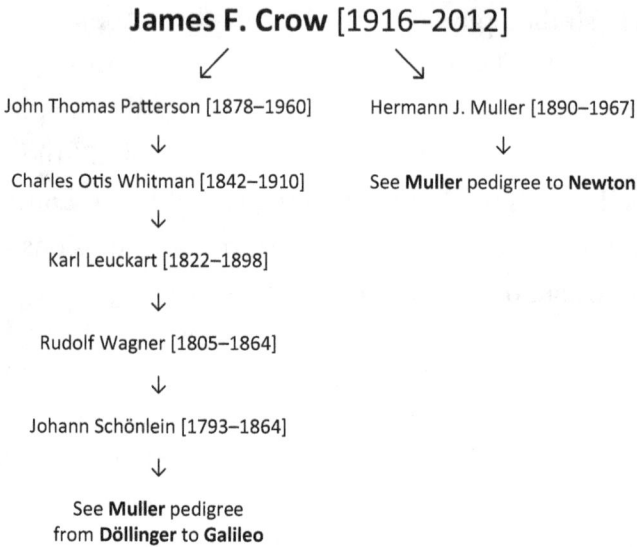

Legend for James Crow Pedigree

James Franklin Crow (1916–2012) was born in Phoenixville, Pennsylvania, and died in Madison, Wisconsin. His father was a college teacher. He obtained his PhD at the University of Texas with **Muller, Patterson**, and **Stone** as his major mentors. He chose population genetics which he studied initially with fruit flies and later in humans. He contributed to the estimation of genetic load in the population, the effects of aging on sperm mutation rates, estimating inbreeding in human populations, and estimating induced mutation rates from radiation exposure. He sympathized with Quaker values but considered himself to be a theological atheist.

J. T. Patterson (1878–1960) was born in Piqua, Ohio, and died in Austin, Texas. He received his PhD from **C. O. Whitman** studying embryology at the University of Chicago. He studied the embryology of armadillos when he came to Texas and showed they produced identical quadruplets. He hired **Muller** to the Texas staff and helped Muller obtain an X-ray machine for his research. Patterson shifted from embryology to the evolution of the genus Drosophila, using cytological approaches.

Charles Otis Whitman (1842–1910) was born in Woodstock, Maine, and died in Worcester, Massachusetts. He grew up on a farm to Adventist parents who were pacifists and did not allow the young Whitman to join the Union forces during the Civil War. Whitman later converted to Unitarianism and taught at a Unitarian prep school, Westford Academy, in Massachusetts. He went to Germany (Leipzig) and got a PhD studying embryology with **Leuckart**. He applied for an opportunity to go to Japan for two years and became "father of Japanese Zoology" by bringing microscopy and German Entwicklungsmechanik to Japan's Imperial University. Whitman returned to the United States and attended Alexander and Louis Agassiz's Biological station (then at Penikese Island). He became professor of zoology at the University of Chicago and studied pigeons, pioneering in ethology (animal behavior), pigeon embryology, and the comparative anatomy of pigeon varieties and species. He tried but could not get other pigeons to nurture the last passenger pigeon eggs.

Karl G. F. R. Leuckart (1822–1898) was born in Helmstedt, Germany, and died in Leipzig, Germany. He was a parasitologist working out life cycles and hosts for human tapeworms, trichinosis, and for sheep liver flukes. His work led to inspections of meat for parasites in industrialized countries.

Rudolf Wagner (1805–1864) was born in Bayreuth, Germany, and died in Göttingen, Germany. He described the germinal vesicle of the ovum in mammalian ovaries. It was later shown to be an arrested stage in prophase I of meiosis and it resumes cell division (when a surge of luteinizing hormone occurs during mature egg formation). He was a vitalist who followed Goethe's view of biology.

Johann L. Schönlein (1793–1864) was born and died in Bamberg, Germany. He broke tradition and lectured in German instead of Latin. He identified the cause of ringworm as a scalp fungal parasite *Trichophyton schönlandii*). He also gave the name tuberculosis to what was then called consumption.

Charles Benedict Davenport [1866–1944]

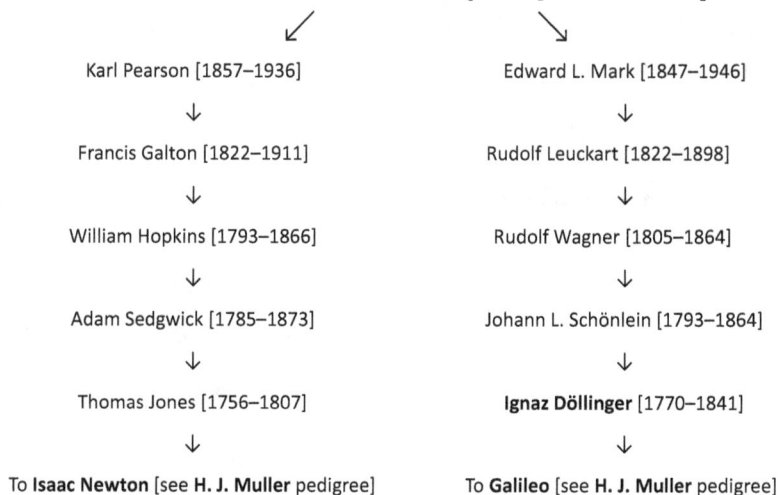

↙	↘
Karl Pearson [1857–1936]	Edward L. Mark [1847–1946]
↓	↓
Francis Galton [1822–1911]	Rudolf Leuckart [1822–1898]
↓	↓
William Hopkins [1793–1866]	Rudolf Wagner [1805–1864]
↓	↓
Adam Sedgwick [1785–1873]	Johann L. Schönlein [1793–1864]
↓	↓
Thomas Jones [1756–1807]	**Ignaz Döllinger** [1770–1841]
↓	↓
To **Isaac Newton** [see **H. J. Muller** pedigree]	To **Galileo** [see **H. J. Muller** pedigree]

Legend for Charles Davenport Pedigree

Note: Davenport's pedigree is like Muller's in following the Mark to Döllinger pathway where it continues into the Galileo branch. He also enters the pathway to Newton through Pearson's mentorship with Francis Galton, who connects Hopkins to the Newton pathway.

Charles Benedict Davenport (1866–1944) was born in Stamford, Connecticut, and died at Cold Spring Harbor, New York. His father was an engineer and real estate broker and he had his son home-schooled to learn disciplined habits of study and work. He started out in engineering but switched after hearing some lectures of **E. L. Mark** at Harvard and got his PhD in 1892. He liked quantitative approaches to biological problems. He corresponded and visited **Karl Pearson** and **Francis Galton**. But in 1901 he shifted from biometrics to the new Mendelism. He became the founding director of the Cold Spring Harbor Laboratory. He extended Mendelism to chickens and humans. His interests in eugenics unfortunately were biased and the Eugenics Record Office he created shifted to political lobbying for compulsory sterilization laws and restrictive immigration laws based on racial and ethnic bias.

Edward Laurens Mark (1847–1946) received his BA from the University of Michigan in 1871 and went to Leipzig, Germany for his PhD in 1876. He studied parasitology with **Rudolf Leuckart** and brought back to Harvard the scientific practices he absorbed in five years in Germany. He studied cell lineages in development and comparative anatomy.

Rudolf Leuckart (1822–1898) was born in Helmstedt and died in Leipzig, both in Germany. He was a parasitologist who worked on the life cycles and hosts of parasites.

Rudolf Wagner (1805–1864) was born in Bayreuth and died in Göttingen, Germany. He wrote a volume on contagious diseases and studied microscopic anatomy, discovering that the ovum was within what he called a germinal vesicle. Prior to that, the vesicle itself was considered the surface of the egg.

Johann Schönlein (1793–1864) was born and died in Bamberg, both in Germany. He stressed medical education with a focus on science as its basis for diagnosis and treatment.

Georges Cuvier (1769–1832) was born in Montbéliard, France, and died in Paris, France. He introduced comparative anatomy, the study of fossils (paleontology), and a theory of extinctions caused by tsunamis and replacements by new creations of life forms. The fossils in each layer of rock suggested this "catastrophe theory." He rejected Lamarck's evolutionary views.

Max Delbrück [1906–1981]

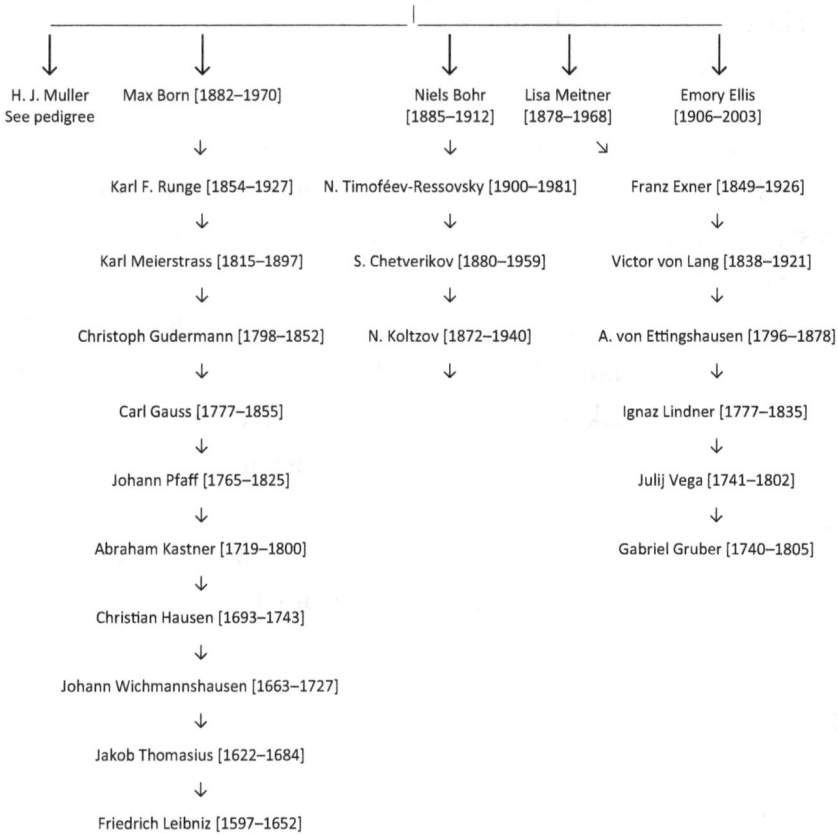

↓	↓	↓	↓	↓
H. J. Muller See pedigree	Max Born [1882–1970]	Niels Bohr [1885–1912]	Lisa Meitner [1878–1968]	Emory Ellis [1906–2003]

Karl F. Runge [1854–1927]   N. Timoféev-Ressovsky [1900–1981]   Franz Exner [1849–1926]

↓ ↓ ↓

Karl Meierstrass [1815–1897]   S. Chetverikov [1880–1959]   Victor von Lang [1838–1921]

↓ ↓ ↓

Christoph Gudermann [1798–1852]   N. Koltzov [1872–1940]   A. von Ettingshausen [1796–1878]

↓ ↓ ↓

Carl Gauss [1777–1855]   Ignaz Lindner [1777–1835]

↓ ↓

Johann Pfaff [1765–1825]   Julij Vega [1741–1802]

↓ ↓

Abraham Kastner [1719–1800]   Gabriel Gruber [1740–1805]

↓

Christian Hausen [1693–1743]

↓

Johann Wichmannshausen [1663–1727]

↓

Jakob Thomasius [1622–1684]

↓

Friedrich Leibniz [1597–1652]

Legend for Pedigree of Max Delbrück

Max Delbrück (1906–1981) was born in Berlin, Germany, and died in Pasadena, California. His father was a historian and he was the great-great-grandson of **Justus von Liebig**. He became a physicist and got his PhD in 1930 at Göttingen. In 1932 he joined discussions on the gene at the Kaiser Wilhelm Institute with **Timoféev-Ressovsky, Zimmer**, and **Muller**. He left Germany in 1937 and came to the United States where he worked at Vanderbilt University and at Caltech. While at Caltech he did experiments with **Ellis** showing bacteriophage did not multiply exponentially, they emerged in single steps. He founded the phage group with **Luria** and **Hershey**, sharing a Nobel Prize for their work in establishing molecular genetics.

Max Born (1882–1970) was born in Breslau (now Wroclaw, Poland), Germany, and died in Göttingen, Germany. His father was a professor of embryology. His mother died when he was 7 years old. His father was Jewish but converted to Lutheranism. Born received his BA in 1901 and studied mathematics with **Klein, Hilbert**, and **Minkowski**. He got his PhD in mathematics in 1906 and took an interest in applying mathematics to physics. He related relativity to electrodynamics using mathematical approaches. In 1932 he was forced to leave Germany and worked in Edinburgh. During the World War I he studied artillery efficiency. He was awarded a Nobel Prize for his contributions to quantum mechanics.

Karl Runge (1854–1927) was born in Bremen and died in Göttingen, both in Germany. His father was a consul in Havana. He received his PhD in mathematics in Berlin in 1880. He worked in astrophysics and used spectral lines of stars and compared them to spectral lines of elements. This allowed astronomers to study atomic distribution in stars and study stellar evolution. He was a pioneer in applied mathematics.

Karl Weierstrass (1815–1897) was born in Ennigerloh and died in Berlin, both in Germany. His father was a government officer. He chose studying mathematics at the Munster Academy. He worked on proofs of the calculus of variation. He taught mathematics and **Husserl** and **Cantor** were his most distinguished students.

Christoph Gudermann (1798–1852) was born in Vienenburg, Germany, and died in Munster, Germany. His father was a school teacher. He worked on the mathematics of elliptical functions and spherical geometry.

Carl Friedrich Gauss (1777–1855) was born in Brunswick and died in Göttingen, both in Germany. His parents were working class and his mother was illiterate. He was a child prodigy and word of his abilities allowed the Duke of Brunswick to pay for his education. He made contributions to numerous branches of mathematics. He applied mathematics to optics for determining the curvature of lenses. He applied sampling to demonstrate random distributions producing normal or bell or Gaussian curves. He was considered the greatest mathematician since antiquity.

Abraham Kastner (1719–1800) was born in Leipzig and died in Göttingen, both in Germany. His father was a professor of law. Kastner became a math and physics professor at Göttingen. He wrote texts in mathematics and a history of mathematics.

Christian August Hausen (1693–1743) was born in Dresden and died in Leipzig, both in Germany. He was a mathematician and a physicist. He studied static electricity and wrote a book on his experiments.

Johann Christoph Wichmannshausen (1663–1727) was born in Ilsenberg and died in Wittenberg, both in Germany. He was a philologist and taught Hebrew and Middle East languages as well as philosophy.

Jacob Thomasius (1622–1684) was born and died in Leipzig, Germany. He was a philosopher and wrote a criticism of Spinoza's views of the universe as a pantheistic system.

Friedrich Leibniz (1597–1652) was born in Altenburg, Germany, and died in Leipzig, Germany. He taught moral philosophy and was an actuary. He is also the father of the polymath **Gottfried Leibniz**.

Niels Bohr (1885–1962) was born in Copenhagen, Denmark, and died in Carlsberg, Denmark. His mother was Jewish and father Lutheran, but Bohr identified himself as an atheist. He won a Nobel Prize for his work on the structure of atoms, demonstrating that electrons shift orbits and atomic components sometimes behave as a stream of particles and other times as waves. He called these apparently contradictory behaviors by the term complementarity. In 1943 he escaped Nazi arrest by going to Sweden and then to Scotland. He resisted German and Russian efforts to have him participate in the development of nuclear weapons. Both British and American intelligence considered him unreliable as a potential contributor to their atomic weapons programs.

Nicolay Timoféev-Ressovsky (1900–1981) was born in Moscow, Russia, and died in Oblinsk, Russia. He received his PhD with **Koltzov** and studied genetics. He liked **Muller's** induction of mutations with X-rays and studied rates of mutation and then designed (with **Max Delbrück** and **Zimmer**) a physical approach to gene size, using target theory. It was published as "On the nature of gene mutation" in 1933 and is sometimes identified as the first paper in molecular genetics. Timoféef-Ressovsky worked at the Kaiser Wilhelm Institute in Berlin and after the war was arrested and sent to face trial. He was spared execution because his son was killed fighting against Nazis and Timofeef-Ressovsky was needed for the radiobiology program as an adjunct to the USSR effort to develop nuclear weapons. He was sent to Siberia and did radiation genetic research until his retirement.

Sergei Chetverikov (1880–1959) was born in Moscow, Russia, and died in Nizhny-Novgorod, USSR. He applied classical genetics to Darwinian evolution, but his paper was in Russian and few knew of his contribution to what would later be called the New Synthesis. He was arrested twice for political activism. In 1929 the OGPU considered him unreliable but he was released. In 1948 Lysenko's published articles falsely denouncing him for fascist genetic views and he was dismissed from his Institute at Gorky University.

Nicolay Koltzov (1872–1941) was born in Moscow and died in Saint Petersburg, both in Russia. His father was an accountant. His PhD in 1905 was on cellular structure and he introduced the term cytoskeleton. He predicted the gene would be a large molecule with two complementary parts. He was arrested in 1920 but saved by intervention of Maxim Gorky. In 1939 he was denounced by Lysenkoists and died, probably by poisoning by the NKVD.

Emory Ellis (1906–2003) was born in Grayville, Illinois, and died in Santa Maria, California. He received his PhD in biochemistry at Caltech in 1934 and chose viruses to study as possible agents to kill tumor cells. **Delbrück**, who was visiting at Caltech liked Ellis's bacteriophages and they used the T series of bacteriophage that infected *E. coli*. In 1939 they published "Growth of bacteriophage" and launched molecular genetics. They showed the virus did not multiply exponentially but produced single bursts from infected cells, each cell releasing dozens of mature viruses. Ellis left phage genetics and returned to cancer biology and space biology for the rest of his career.

Franz Exner (1849–1926) was born and died in Vienna, Austria. His father was a professor of philosophy. Exner chose physics and chemistry as his field getting his PhD in 1871. He studied the diffusion of molecules through membranes and water layers. He mentored dozens of graduate students including **Erwin Schrödinger** and **Lise Meitner**. His wife was active in local affairs and Exner championed women in higher education and formed a group with **Ernst Mach** and **Victor von Lang** to bring more women to the University of Vienna.

Victor von Lang (1838–1921) was born and died in Vienna. He worked on the physics of crystals studying isomeric forms of crystals. He spent part of his education with Faraday in London.

Milislav Demerec [1895–1966]

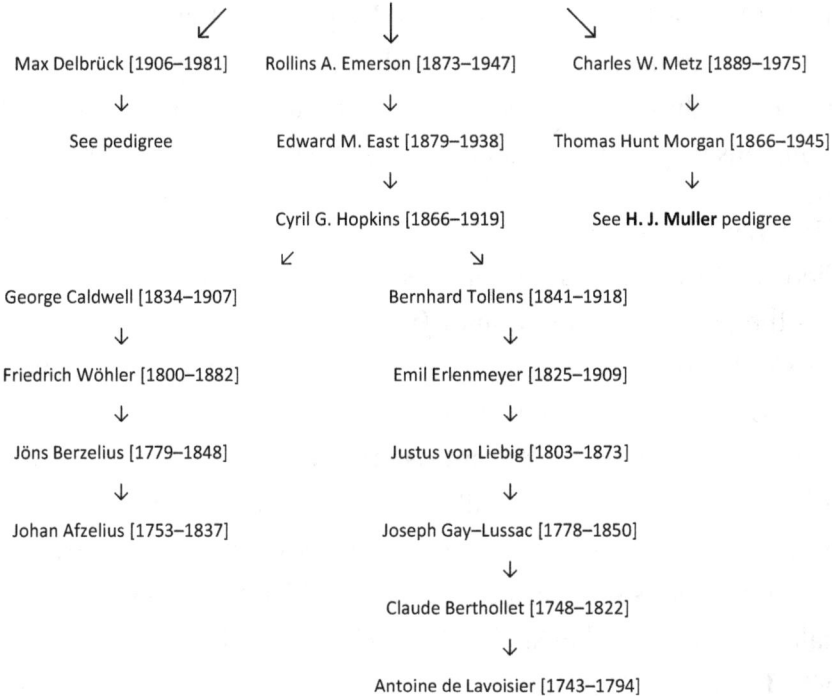

↙	↓	↘
Max Delbrück [1906–1981]	Rollins A. Emerson [1873–1947]	Charles W. Metz [1889–1975]
↓	↓	↓
See pedigree	Edward M. East [1879–1938]	Thomas Hunt Morgan [1866–1945]
	↓	↓
	Cyril G. Hopkins [1866–1919]	See **H. J. Muller** pedigree

↙	↘
George Caldwell [1834–1907]	Bernhard Tollens [1841–1918]
↓	↓
Friedrich Wöhler [1800–1882]	Emil Erlenmeyer [1825–1909]
↓	↓
Jöns Berzelius [1779–1848]	Justus von Liebig [1803–1873]
↓	↓
Johan Afzelius [1753–1837]	Joseph Gay–Lussac [1778–1850]
	↓
	Claude Berthollet [1748–1822]
	↓
	Antoine de Lavoisier [1743–1794]

Milislav Demerec (1895–1966) was born in Kostajnike, Croatia, and died in Laurel Hollow, New York. He got his BA at Krizevci University in 1916 and emigrated to the United States in 1919. He got his PhD at Cornell in 1923 studying maize genetics with **Emerson**. He switched to fruit flies and he did research at Cold Spring Harbor Laboratory studying mutable genes. He founded *Drosophila Information Service* and became Director of the Laboratory after Blakeslee's retirement. He shifted to microbial genetics in the 1940s and used X-rays to produce strains of penicillin that greatly increased penicillin production. He showed genetic fine structure in bacteria while Benzer did so in bacteriophage.

Charles William Metz (1889–1975) was born in Wyoming and died at Woods Hole, Massachusetts. He got his BA at Pomona College in California and his PhD at Columbia University working with **Morgan's** fly lab. He studied *D. virilis* and obtained numerous mutations mapping them for comparison with *D. melanogaster.*

Max Delbrück (1906–1981) was born in Berlin, Germany, and died in Pasadena, California. His father was a historian and he was the great-great-grandson of **Justus von Liebig**. He became a physicist and got his PhD in 1930 at Göttingen. In 1932 he joined discussions on the gene at the Kaiser Wilhelm Institute with **Timoféev-Ressovsky, Zimmer, and Muller**. He left Germany in 1937 and came to the United States where he worked at Vanderbilt University and at Caltech. While at Caltech he did experiments with **Ellis** showing bacteriophage did not multiply exponentially, they emerged in single steps. He founded the phage group with **Luria** and **Hershey**, sharing a Nobel Prize for their work in establishing molecular genetics.

Rollins Emerson (1873–1947) was born in Pillar, New York, and died in Ithaca, New York. At age 7 his family moved to Nebraska. He liked agriculture and attended the University of Nebraska for a BA in 1897 and studied beans, confirming the Mendelian findings for many different traits. He then went to Harvard to study maize with **E. M. East** for his PhD. He joined Cornell's Botany Department and studied gene mutations in maize, making Cornell a world center for plant genetics.

Edward Murray East (1879–1938) was born in Du Quoin, Illinois, and died in Boston, Massachusetts. His father was a mechanical engineer. He worked on hybrid corn and showed dominance was the major reason for hybrid vigor after two inbred strains of corn were crossed. He applied this to the production of hybrid corn which vastly increased yields on farms. He also applied this to his views on eugenics. He felt inbreeding was bad if there were recessive lethal or detrimental genes present. But if there were no such genes, inbreeding could be used to produce strains of desired corn or other plants. He also applied this to the growing and later discredited eugenics movement and his views were racist and elitist. Whites, he claimed, were superior and blacks were inferior. He also advocated eugenics favoring sterilization of social failures (paupers, psychotics, or the mentally deficient).

Cyril George Hopkins (1866–1919) was born in Chatfield, California, and died in Gibraltar. He got his PhD at Cornell in 1898 studying agricultural chemistry. His dissertation was on "The chemistry of a corn kernel." He taught agricultural chemistry at the University of Illinois, showing how trace minerals and fertilizers could keep soil from being exhausted. In 1919 he wanted to help Greece recover from a war that was accompanied by crop failures. He was regarded as a hero for his contributions to that recovery, but he caught malaria and died in Gibraltar.

George Chapman Caldwell (1834–1907) was born in Framingham, Massachusetts, and died in Ithaca, New York. He received his BA at Harvard and then went to Germany to study with **Robert Bunsen** and **Friedrich Wöhler**. He returned with a PhD and during the Civil War served on the Sanitary Commission. He went to Cornell as its first appointed professor and built the chemistry department offering programs in both basic chemistry and agricultural chemistry.

Friedrich Wöhler (1800–1882) was born in Frankfurt, Germany, and died in Göttingen, Germany. He studied with **Berzelius** and **Gmelin**. He obtained urea from ammonium cyanide and this was the first organic compound that was produced from inorganic components, establishing the existence of organic chemistry. Many chemists then believed organic compounds depended on a vital principle for their synthesis. Wöhler also discovered aluminum, titanium, yttrium, and beryllium as elements. He believed compounds could be broken down into their elements by appropriate chemical means.

Jöns Jacob Berzelius (1779–1848) was born in Östergötland, Sweden, and died in Stockholm, Sweden. His father was a school teacher but died when Jöns was one year old. His mother died when he was 7 years old. He was raised by relatives. He began his education at Uppsala University hoping to be a physician but switched to chemistry and physics. He showed that compounds could be separated by electric current into charged positive and negative ions. He determined the atomic weights of elements. He found silicon, selenium, thorium, and cerium. He also provided the nomenclature symbolism for elements and compounds, 1 or 2 letters and a subscript for element in a compound (thus CO_2 is carbon dioxide; $CaCl_2$ is calcium dichloride).

Johan Afzelius (1753–1837) was born in Larv, Sweden, and died in Uppsala, Sweden. He got his PhD with **Olof Bergman**. He isolated formic acid from ants and showed it is the cause of its stinging bite. He isolated and named oxalic acid.

Bernhard Tollens (1841–1918) was born in Hamburg, Germany, and died in Göttingen, Germany. He studied with **Karl Mobius** as an undergraduate in 1857 and with **Friedrich Wöhler** in 1864 for his PhD. He spent his career working on carbohydrates, isolating and characterizing the sugars found in them.

Emil Erlenmeyer (1825–1909) was born in Taunusstein, Germany, and died in Aschaffenburg, Germany. His father was a Protestant minister. As an undergraduate he heard von Liebig's lectures and decided he wanted to be a chemist. He studied with **Bunsen** and **von Liebig** and received his PhD in 1850. He was the first to use double and triple bonds in his chemical formulae for carbon atoms. He invented the Erlenmeyer flask. He analyzed alcohols and aldehydes and their composition.

Justus von Liebig (1803–1873) was born in Darmstadt, Germany and died in Munich, Germany. His father was a hardware merchant. He also experienced the hardships of 1816 (the year without a summer) and he devoted his life to help humanity cope with natural catastrophes. He was a founder of organic chemistry and of the chemical fertilizer industry. He invented the bouillon cube. He provided the first applied chemical laboratory and had over 700 students who took that program. He demonstrated compounds had side groups that acted like elements in their combining properties.

Joseph Gay-Lussac (1778–1850) was born in Sant-Leonard-de-Noblat, France, and died in Paris, France. His father was a lawyer. He was asked during the French Revolution to start a technical school in Paris. He worked out the relation between gas volumes at different temperatures and pressures and showed the resulting volumes were mathematically predictable. He worked out the amount of nitrogen and oxygen in different compounds. He discovered the element boron.

Claude Berthollet (1748–1822) was born in Talloires, France, and died in Arceuil, France. He obtained a medical degree in Turin, Italy, and studied dyes and bleaches including sodium hypochlorite. He showed the composition of ammonia (NH_3).

Antoine de Lavoisier (1743–1794) was born and died in Paris, France. He was of a noble family and his father was a court lawyer. He chose not to follow his father's career although he got a law degree and used his wealth to support his own research in chemistry. He showed combustion requires oxygen and that combustible material gained weight after combustion with air. He discovered oxygen as a gas in the air. He introduced the suffixes used in chemistry (-ic, -ate, -ite, -ous, etc.). He was arrested for his role as a tax collector for the King and was beheaded.

Hugo De Vries [1848–1935]

✓   ↓

Wilhelm Hofmeister [1824–1877]   Julius von Sachs [1832–1897]

↓

Jan Purkinje [1787–1869]

↓

Bernard Bolzano [1781–1848]

↓

Franz von Gerstner [1756–1832]

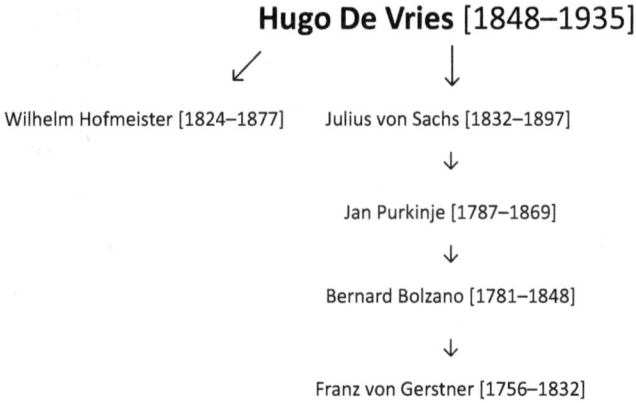

Legend for Hugo de Vries

Hugo de Vries (1848–1935) was born in Haarlem and died in Lunteren, both in Holland. His father was a minister of justice and later Prime Minister of the Netherlands. He took an interest in botany and plant physiology. He studied with **Hofmeister** and **von Sachs** in Leipzig, Germany and worked out the physiology and physical chemistry of osmosis in plant cells and tissues. He taught at the University of Amsterdam for 41 years and shifted to problems of heredity. His 1889 book, *Intracellular Pangenesis*, shifted Darwin's gemmules from circulating hereditary units to fixed units within the nuclei of cells. He believed changes in number, location, and function of such units could interpret how heredity worked at the cellular level. Darwin called them gemmules and de Vries called them pangenes. De Vries was a co-discoverer of Mendel's findings of fixed ratios for assortment and independent assortment. His "Mutation Theory" was a mixture of good insights and events, later interpreted as polyploidy, aneuploidy, and chromosome rearrangements for new species of primroses that he identified.

Wilhelm Hofmeister (1824–1877) was born in Leipzig and died in Lindenau, Germany. His father was a publisher of scientific books and Hofmeister became an autodidact, doing research and publishing his findings in his father's business. He discovered and interpreted the alternation of generations in many plants that have separate gametophyte and sporophyte generations in their life cycles. He did not accept an academic position until later in his life at age 39.

Julius von Sachs (1832–1897) was born in Leipzig and died in Würzburg, both in Germany. His father was an engraver. At 16, his father died and the following year his mother and a brother died of an epidemic. He was adopted by his teacher, **Jan Purkinje** with whom he studied plant physiology. He worked out the role of photosynthesis in plants, demonstrating by experiment that leaves exposed to sunlight produced starch granules.

Jan Purkinje (1787–1869) was born in Libochovice and died in Prague, both in Czechoslovakia. His father was an estate manager. Purkinje chose medicine as a career and received his MD in 1818 at Prague's Charles University. He was the first to use a microtome to slice tissues and prepare slides of their tissues. He identified sweat glands in skin, and he classified fingerprints by their major patterns.

Bernard Bolzano (1781–1848) was born and died in Prague, Czechoslovakia. He taught philosophy and mathematics.

Franz von Gerstner (1756–1832) was born in Komoto, Czechoslovakia, and died in Mladejov, Austria. He taught mathematics and physics. His major finding was the theory of waves and the mathematics of their movements.

Theodosius Dobzhansky [1900 –1975]

↙   ↘

Yuri Filipchenko [1882–1930]   Alfred H. Sturtevant [1891–1970]

↓   ↓

Theodor Boveri [1862–1915]   Thomas Hunt Morgan [1866–1945]

↓   ↓

See pedigree of **Boveri**   See **H. J. Muller** pedigree

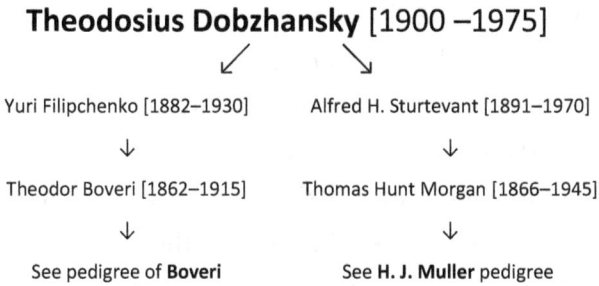

Legend for Pedigree of Theodosius Dobzhansky

Theodosius Dobzhansky (1900–1975) was born in Nemerov, Ukraine, and died in San Jacinto, California. His father was a high school mathematics teacher. Dobzhansky went to St. Petersburg to study with Filipchenko who recommended he do graduate work at Columbia University with **Morgan**. He did his dissertation with **Sturtevant**. He then stayed with the fly lab when it moved to Caltech and wrote *Genetics and the Origin of Species* which became a classic. His research work was on chromosome rearrangements associated with speciation. He called these and other factors, isolating mechanisms. Dobzhansky taught at Columbia University and at the Rockefeller Institute before retiring to California. He died of leukemia.

Yuri Filipchenko (1882–1930) was born in Zlyn, Russia, and died in Leningrad, USSR. His father was a farmer. He read Darwin's *Origin of Species* and **Nägeli's** monograph on heredity while in high school and wanted to be a geneticist. He had been arrested in the 1905 Revolution and after the Bolshevik Revolution he founded the first genetics department in Russia. He coined the terms microevolution and macroevolution, using the first term for Darwinian speciation and the second term for the origin of body plans which he felt had a different evolutionary origin.

Theodor Boveri (1862–1915) was born in Bamberg and died in Würzburg, Germany. He obtained his MD in Munich and studied cell division by using experimental approaches. He agitated sea urchin fertilized eggs and caused abnormal chromosome numbers in them. The larvae either failed to develop or produced abnormal embryos. This led Boveri to propose the chromosome theory of heredity. His work both stimulated and supported the work of **E. B. Wilson** and his students in promoting that theory.

Hans Driesch [1867–1941]

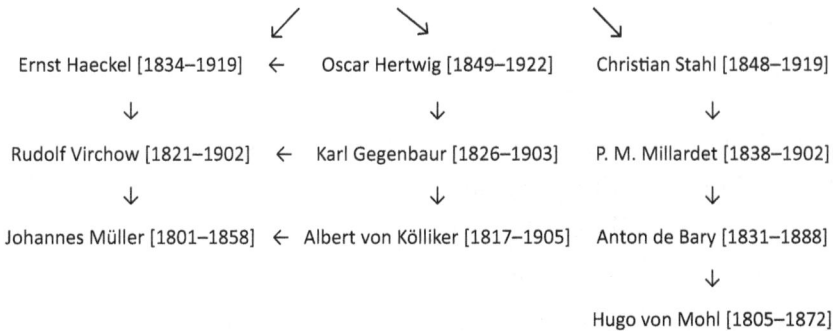

↙   ↘   ↘

Ernst Haeckel [1834–1919]   ←   Oscar Hertwig [1849–1922]   Christian Stahl [1848–1919]

↓   ↓   ↓

Rudolf Virchow [1821–1902]   ←   Karl Gegenbaur [1826–1903]   P. M. Millardet [1838–1902]

↓   ↓   ↓

Johannes Müller [1801–1858]   ←   Albert von Kölliker [1817–1905]   Anton de Bary [1831–1888]

↓

Hugo von Mohl [1805–1872]

Legend for Intellectual Pedigree of Hans Driesch

Hans Driesch (1867–1941) was born in Bad Kreuznach and died in Leipzig, both in Germany. He got his PhD in 1889 in Munich and went to Naples Marine Biological Station to study and do research. He was an early founder of Entwicklungsmechanik or experimental embryology. He taught **Morgan** this approach in Naples. He used sea urchins and separated blastomeres at the 2 or 4 cell stage and showed that each produced an identical twin of normal, but smaller embryo. This implied that the design for the organism was in each cell of the blastomeres. He called this totipotency. He also thought this proved vitalism and life could not be reduced to mechanistic explanation at the cellular level. He also switched fields and taught philosophy.

Ernst Haeckel (1934–1919) was born in Potsdam, Germany. He obtained his MD at Würzburg and Berlin, Germany. He got his PhD with **Gegenbaur**. He studied radiolarians and named numerous species whose life cycles he studied and published. He prepared the first phylogenetic tree for the evolution of life on earth. He promoted Darwinian evolution by writing best sellers for the public. He taught at Jena for 47 years. He is best known for his recapitulation theory: Ontogeny recapitulates phylogeny (the sequence of embryonic development form fertilization to adult reflects the past ancestry of life on earth seen in the fossil record.

Rudolph Virchow (1821–1902) was born in Schivelbein and died in Berlin, both in Germany. He was sent in 1848 to study the high mortality and morbid rate among miners in Silesia (southeast Germany). He claimed it was due to low wages. He was fired and went to south-western Germany and studied pathology and founded the field of cellular pathology. This included the single cell origin of cancer. He confirmed **Remak's** view that cell division requires prior cells for formation, an action Virchow designated as the cell doctrine "all cells arise from pre-existing cells". He championed hygiene programs in Germany and free health examination of children. He was a foe of Darwinian evolution. He rejected an Aryan ancestry of Germans and showed that Aryan traits (blond, blue eyed) were a minority of the German population.

Johannes Müller (1801–1858) was born in Coblenz and died in Berlin, both in Germany. His father was a shoemaker and the young Müller considered the priesthood to get an education. He switched to medicine at Bonn when he was 18 and studied the physiology of sense organs, using tools from chemistry and physics to put physiology on an experimental level. His text in physiology was a much-used classic in the nineteenth century. He mentored numerous students who applied his experimental approach to the life sciences.

Oscar Hertwig (1849–1902) was born in Friedberg and died in Berlin, both in Germany. He was the older brother of **Richard Hertwig**. Richard taught at Munich and Oscar at Berlin. Oscar

Hartwig's research was on the maturation of reproductive cells. He showed in 1876, that one sperm enters and egg and fertilizes the egg nucleus. He speculated that nucleic acids (found in abundance in a sperm head) may be the hereditary material. He rejected the role of chance in evolution.

Karl Gegenbaur (1829–1903) was born in Würzburg and died in Heidelberg, both in Germany. He studied comparative anatomy and emphasized homologous relations of embryonic structures, such as the bone formation of hands, paws, fins, and wings. He showed the skull is not derived from anterior bones of the spinal column but from separate bones of the skull.

Edward M. East [1879–1938]

↓

Cyril Hopkins [1866–1919]

↓

Bernhard Tollens [1847–1918]

↙   ↘

| Friedrich Wöhler [1800–1882] | E. Erlenmeyer [1825–1903] |

↙   ↓   ↓

| Leopold Gmelin [1788–1853] | Jöns Berzelius [1779-1848] | Justus von Liebig [1803–1873] |

↓   ↓

| Anders Ekeberg [1767–1813] | Joseph Gay-Lussac [1788–1850] |

↓   ↓

| Carl Thunberg [1743–1828] | Claude Berthollet [1748–1856] |

↙   ↘   ↓

Carl Linnaeus [1707–1778]   Antoine de Lavoisier [1743–17984]

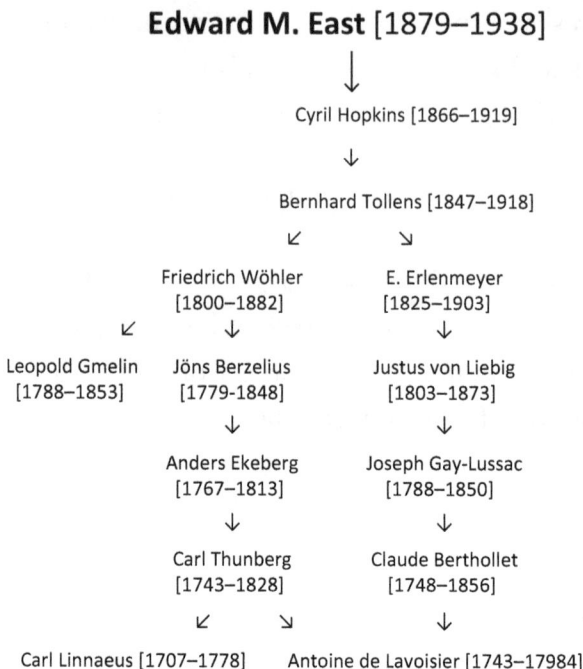

Legend for Edward Murray East

Edward Murray East (1879–1938) was born in Du Quoin, Illinois, and died in Boston, Massachusetts. His father was a mechanical engineer. East got his PhD in 1907 at the University of Illinois. He began in plant chemistry and shifted to plant breeding discovering multiple factor inheritance in maize. He shifted to a study of heterosis or hybridization and its consequences, applying this to the development of hybrid corn. His book *Inbreeding and Outbreeding* was a best seller and he became an advocate for the American eugenics movement favoring restrictive immigration laws and compulsory sterilization of those deemed medically or psychologically unfit to reproduce. His views were racist, sexist, and based on bias.

Cyril G. Hopkins (1866–1919) was born in Chatfield, Minnesota, and died in Gibraltar, Africa. He went to Göttingen for his PhD in 1899 in soil chemistry and contributed to crop rotation and use of fertilizers in the soil to aid farmers to improve crop yields. In 1898 he got his PhD at Cornell. He went to Greece to help their farmers use the methods he advocated to replenish their soil but on his return trip caught malaria and died in Gibraltar.

Bernhard Tollens (1841–1918) was born in Hamburg and died in Göttingen, Germany. He worked out the sugar contained in carbohydrates such as starch.

Friedrich Wöhler (1800–1882) was born in Eschersheim and died in Göttingen in Germany. He is best known for synthesizing urea and other organic chemicals, founding the field of organic chemistry.

Leopold Gmelin (1788–1853) was born in Göttingen and died in Heidelberg. He was the first to identify hydrochloric acid as the major agent in the stomach to begin the digestion of food. He isolated and named cholesterol.

Jöns Berzelius (1779–1848) was born in Östergötland and died in Stockholm, Sweden. He discovered the elements thorium, silicon, cesium, and selenium. He contributed to the symbolism used for chemical reactions.

Anders Ekeberg (1767–1813) was born in Karlskrona and died in Uppsala, Sweden. He discovered tantalum. His career was limited by an early onset deafness.

Carl Thunberg (1743–1828) was born in Jonköping and died in Fienberg, Sweden. He went on several expeditions to collect specimens for Linnaeus. He especially explored the East Indies, Japan, and Africa.

Carl Linnaeus (1707–1778) was born in Rashult, Sweden, and died in Hammeby, Sweden. His father was a Lutheran minister. He made four voyages to northern Sweden and Lapland and came back describing the flora and fauna (as well as Lapp culture) he observed. This led to his *Systema Natura* and the binomial classification of species starting in 1753 with many editions for the rest of his career.

E. Erlenmeyer (1825–1903) was born in Taunusstein, Germany, and died in Aschaffensburg, Germany. He studied the chemistry of alcohols, aldehydes, and introduced the use of double and triple carbon bonds to molecules bearing carbon.

Justus von Liebig (1803–1873) was born in Darmstadt and died in Munich in Germany. His father sold paint and varnish and Liebig chose a chemical career. He introduced nitrate fertilizers in soils, and he studied the chemistry of foods and the first synthetic foods (bouillon cubes).

Joseph Gay-Lussac (1778–1850) was born in Saint Lenard de Noblett and died in Paris, France. He showed that water consisted of two parts hydrogen gas to one-part oxygen gas. He also devised the measure of alcohol in wines and liquors (called "proofs").

Claude Berthollet (1748–1856) was born in Talloires, France, and died in Arceuil, France. He studied the properties of chlorine and made the first commercial bleaches from sodium hypochlorite. He also worked out the atomic composition of ammonia.

Antoine de Lavoisier (1743–1794) was born and died in Paris. He is a founder of modern chemistry, a co-discoverer of oxygen, discoverer of hydrogen, and the first to introduce a usable chemical nomenclature and abbreviations. He was guillotined for his role as a tax collector for the King during the French Revolution.

Ronald Aylmer Fisher [1890–1962]

↙   ↘

James H. Jeans [1877–1946]   Frederick J. M. Stratton [1881–1960]

↓   ↓

Gilbert T. Walker [1868–1958]   H. F. Newall [1857–1943]

↓

John J. Thomson [1856–1940]

↙   ↘

John Strutt (Baron Rayleigh) [1842–1919]   Edward Routh [1831–1907]

↓

Augustus De Morgan [1806–1871]

↓

William Whewell [1794–1866]

↓

John Gough [1757–1825]

↓

John Dalton [1766–1844]

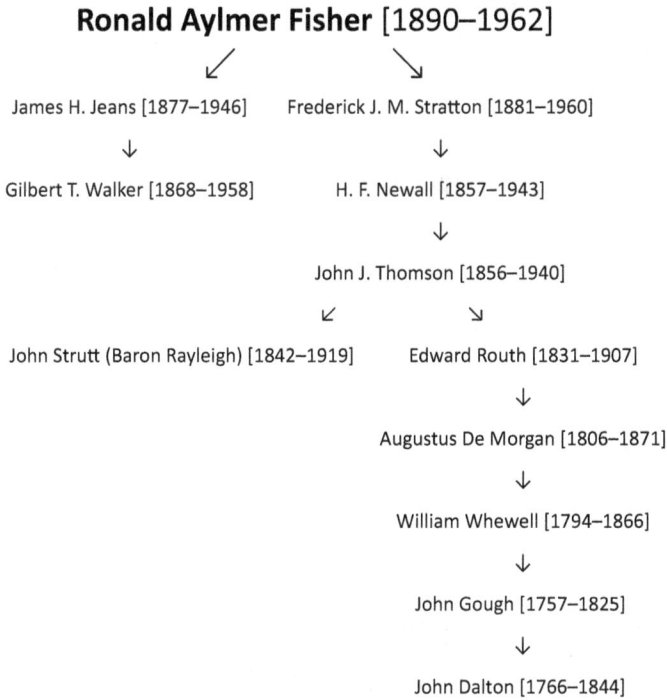

Legend for Intellectual Pedigree for Ronald A. Fisher

Ronald Aylmer Fisher (1890–1962) was born in London, England, and died in Adelaide, Australia. His father was an art auctioneer. He had a gift for mathematics and was largely an autodidact for other fields in which he applied his statistical theorems. He is a founder of population genetics and a major contributor to the New Synthesis that combined Darwinian evolution with population genetics, Mendelism, paleontology, and cytogenetics. He was also a champion of Galtonian eugenics. He was partially blind which forced him to conceive mathematical equations in geometrical imagery.

James Hopgood Jeans (1877–1946) was born in Southampton and died in Dorking, both in England. He applied his knowledge of physics and astronomy to stellar evolution and proposed a steady state model of the universe which was displaced by the Big Bang model of the universe.

Gilbert T. Walker (1868–1958) was born in Rochdale, England, and died in Coulsdon, England. His father was an engineer. He contributed to our understanding of meteor showers and to predictions of weather patterns using statistical approaches to meteorology and astronomic events.

Frederick J. M. Stratton (1881–1960) was born in Edgbaer, England, and died in Cambridge, England. His father was a music critic. He took an interest in astrophysics and served in WWI as commander of the Signal Corps. His specialty was solar physics. He was also a strong supporter of psychical research and served as President of the Society for Psychical Research.

Hugh Frank Newall (1857–1943) was born in Gateshead, England, and died in Cambridge, England. His father was inventor of wire cable and his company provided the cable for the Atlantic Cable. He also provided his son with a telescope that launched his career in astrophysics. He used spectroscopic analysis of stellar spectrographs to determine stellar properties and composition.

John J. Thomson (1856–1940) was born in Manchester and died in Cambridge, both in England. He is best known for discovery of the electron in atoms. He also isolated isotopes of non-radioactive elements. In addition to his own Nobel Prize, eight of his students, six in physics and two in chemistry, went on to win Nobel Prizes.

John Strutt [Baron Rayleigh] (1842–1919) was born in Masdon, England, and died in Witham, England. He won a Nobel Prize for isolating and describing the element Argon, a gas, using gas density as a means of separating gases.

Edward Routh (1831–1907) was born in Quebec, Canada, and died in Cambridge, England. He made his fame as a coach for the Tripos mathematical standings and coached 600 students of whom 28 were senior wranglers. It is the highest award in mathematics at Cambridge. Many applied their mathematics to physics, chemistry, and other sciences.

Augustus De Morgan (1806–1871) was born in Madurai, India, while his father was an officer in the East India Company. Augustus was infected and lost an eye to a childhood disease. His father died when Augustus was 10 years old. He was a contributor to the analysis of logic and mathematics, generating numerous theorems. He was an atheist, and this limited his teaching opportunities. He also believed psychical research was important to distinguish spurious and legitimate research.

William Whewell (1794–1866) was born in Cambridge, England, and died in Lancaster, England. His father was a carpenter. He wrote extensively on the history of science and coined many terms: scientist, physicist, linguistics, catastrophism, uniformitarianism, electrode, cathode, anode, and ion.

John Gough (1757–1825) was born and died in Kendal, England. His father was a wool dyer. He had smallpox which infected his eyes and was partially blind. He was raised a Quaker. He became a philosopher. He taught **Dalton** Latin and Greek and Dalton taught Gough mathematics.

John Dalton (1766–1844) was born in Eaglesfield, England, and died in Manchester, England. His father was a weaver and raised his family as Quakers. He and his brother both were color blind (later shown by DNA analysis of his eyes that he had deuteranopia). He studied measurement of height of mountains using a barometer. He kept a weather diary with 200,000 entries to show weather patterns. He developed an atomic theory with a proposal that atoms were elements and compounds were composed of two or more atoms. He worked out the weights of 20 elements and analyzed 17 compounds that he reduced to their elements. His circular symbols gave way to the **Berzelius** system chemists adopted.

Rosalind Franklin [1920–1958]

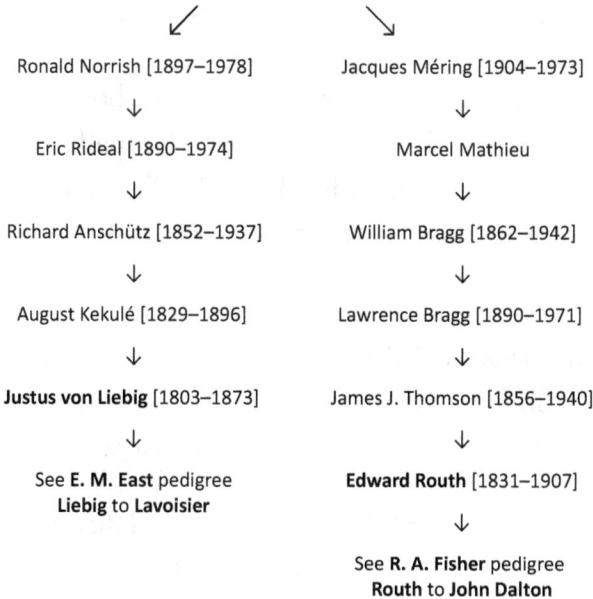

↙ ↘

Ronald Norrish [1897–1978]	Jacques Méring [1904–1973]
↓	↓
Eric Rideal [1890–1974]	Marcel Mathieu
↓	↓
Richard Anschütz [1852–1937]	William Bragg [1862–1942]
↓	↓
August Kekulé [1829–1896]	Lawrence Bragg [1890–1971]
↓	↓
Justus von Liebig [1803–1873]	James J. Thomson [1856–1940]
↓	↓
See **E. M. East** pedigree	Edward Routh [1831–1907]
Liebig to **Lavoisier**	↓
	See **R. A. Fisher** pedigree
	Routh to **John Dalton**

Legend for Rosalind Franklin Pedigree

Rosalind Franklin (1920–1958) was born in Nottingham and died in London, both in England. She studied chemistry with **Ronald Norrish** but found him uninspiring. She worked in France with Jacques **Méring** and learned crystallography from him and studied carbon in coal, graphite, and diamond. She then worked at the University of London and **Randall** hired her to work with (or independently of) **Wilkins** on DNA. This failure by Randall to clarify their roles led to mutual misunderstanding between Franklin and Wilkins on their roles in studying DNA. Their cooperation was minimal and competitive. She returned to England to work with **Bernal** on the crystallography of viruses but at age 37 died of ovarian cancer. Her premature death and the Nobel committee rule of three for an award, excluded her from consideration for a Nobel Prize which went to Watson, Crick, and Wilkins.

Ronald G. W. Norrish (1897–1978) was born and died in Cambridge, England. He was a chemist who made contributions to photochemistry and colloids. His 1924 PhD was on radiation and chemical reactions. These works earned him a Nobel Prize in chemistry.

Eric Rideal (1890–1974) was born in Sydenham, England. He got a PhD in Bonn Germany in 1912 with **Richard Anschütz**. He studied colloids, catalysis, and surface chemistry.

Richard Anschütz (1852–1937) was born and died in Darmstadt, Germany. He was an organic chemist and extended the work of his mentor August Kekulé to other carbon ring molecules.

August Kekulé (1829–1896) was born in Darmstadt, Germany, and died in Bonn, Germany. He was inspired by **Liebig's** lectures to become a chemist. He conceived of the six-carbon ring for the structure of benzene.

Justus von Liebig (1803–1873) was born in Darmstadt, Germany, and died in Munich, Germany. His father was a hardware merchant. He experienced the hardships of 1816 (the year without a summer) causing severe food shortages in central Europe. He was a founder of organic chemistry and the chemical fertilizer industry. He invented the bouillon cube as a supplement to diets. He promoted applied chemistry and taught 700 students how to apply chemistry to industry and agriculture. He was the first to recognize side groups could act like elements in combining with other atoms or compounds.

Jacques Méring (1904–1973) was born in Vilkoviskis, Lithuania, and died in Orleans, France. His father was an engineer. He took an interest in minerology and learned X-ray diffraction to study crystal structure from Marcel Mathieu.

Marcel Mathieu was a French crystallographer who studied with **William Bragg** in the late 1920s and brought the field to French science. He worked on the structure of cellulose and polymers and found a position for **Rosalind Franklin** when she was looking for a place to apply her skills.

William Bragg (1862–1942) was born in Wigton, England, and died in London, England. He and his son, **Lawrence Bragg**, received a Nobel Prize for their development of X-ray crystallography that they applied successfully with photographs of several common crystals.

Joseph John Thomson (1856–1940) was born in Manchester and died in Cambridge, both in England. He discovered the electron and identified it as a component of the atom, thus creating a structural model of the atom in which electrons were embedded in a pudding-like atom.

John Strutt (Baron Rayleigh) (1842–1919) was born in Maldon, England, and died in Witham, England. His father was an engineer. Rayleigh discovered argon and interpreted black body radiation, the physics of sound, and worked out the equations for aerodynamic lift in flight.

Francis Galton [1822–1911]
↓

William Hopkins [1793–1866]

↓

Adam Sedgwick [1785–1873]

↓

See **H. J. Muller** pedigree to **Newton**

Legend for Pedigree of Francis Galton

Francis Galton (1822–1911) was born in Birmingham, England, and died in London, England. He came from a prominent family of dissenters that included the Galtons, Darwins, and Wedgwoods. He was a child prodigy who chose mathematical (mostly statistical) methods to studying fields that became part of psychology, sociology, and anthropology. He coined the term eugenics and argued it should be a moral science of human reproduction. He introduced the correlation into psychological studies. He made contributions to geography, meteorology, and statistics. He used fingerprints for identification. He introduced the regression to the mean as a measure of human population genetics. He published the first daily weather maps.

William Hopkins (1793–1866) was born in Nottinghamshire, England and died in Cambridge, England. He was a mathematician and coached Cambridge students for the honor of being senior wranglers (top mathematics students) at graduation.

Adam Sedgwick (1785–1873) was born and died in Cambridge, England. He was a geologist and theologian who introduced Darwin to geology and went on expeditions in Great Britain with him. He bitterly opposed Darwin's theory of evolution by natural selection.

See **Sedgwick** to **Newton** in **Muller** reference pedigree.

Archibald Garrod [1857–1936]

Frederick G. Hopkins [1861–1947]   William Bateson [1861–1926]

↓   ↓

Thomas Stevenson [1838–1908]   William K. Brooks [1848–1908]

↓

See **H. J. Muller** pedigree to
Galileo

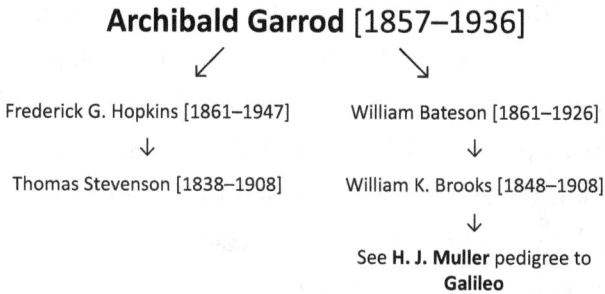

Legend for Pedigree of Archibald Garrod

Archibald Garrod (1857–1936) was born in London and died in Cambridge, both in England. His father was a physician who identified gout as a disease in which uric acid accumulates instead of being degraded into urea. Archibald Garrod studied medicine at Cambridge and took an interest in metabolic diseases that ran in families, usually with cousin marriages. This included alkaptonuria, albinism, cystinuria, and pentosuria. With **Bateson's** help, he showed these were all recessive disorders following Mendelian inheritance. He called them inborn errors of metabolism.

Frederick Gowland Hopkins (1861–1947) was born in Eastbourne, England, and died in Cambridge, England. He was a biochemist who received a Nobel Prize for his discovery of vitamins as necessary nutrients.

Thomas Stevenson (1838–1908) was born in Rainton, England, and died in London, England. He was a toxicologist and helped establish forensic medicine by testifying at poison trials.

William Bateson (1861–1926) was born in Whitby, England, and died in Merton, England. Bateson's father was a college Master. Bateson enjoyed the natural sciences and after taking his MA at Cambridge, he went to Baltimore to study with **William Keith Brooks** at Johns Hopkins University. There he shifted from embryology to heredity and returned to write a book on variation that introduced meristic (duplication of parts) and homeotic (displacement of organs) variations. He felt these were fundamental in the origin of body plans and the evolution of higher taxonomic categories. He was an ardent Mendelian, confirming Mendelism and extending it to epistasis (the interaction of two or more genes in the formation of a character. Bateson called the new field, Genetics, in 1906.

William Keith Brooks (1848–1908) was born in Cleveland, Ohio, and died in Baltimore, Maryland. He received his PhD from Harvard studying with **Louis** and **Alexander Agassiz**. At Johns Hopkins he studied tunicates and oysters, stressing their embryology as clues to the evolution of classes and phyla. He advocated working with living organisms and working out life cycles. He was Lamarckian in his evolutionary views but believed heredity lacked experimental approaches and needed the attention of scientists.

Karl Gegenbaur [1826–1903]
↓

↓	↓	↓	↓
	Rudolph Virchow	Heinrich Müller	Franz Leydig
	↓	↓	↓
Friedrich G. Henle →	Johannes Müller	Ignaz Döllinger	Martin Münz
↙	↓	↓	
↙	↓	See **Muller** pedigree	
August Mayer	Karl Rudolphi		
↓	↓		
Carl Kielmeyer	Christian Weigle		
↓	↓		
Johann Blumenbach	Johann Erxleben		
↓	↓		
Christian Büttner	Abraham Kastner		
↓	↓		
Herman Boerhaave	Johann Wichmannshausen		
	↓		
	Otto Menke		
	↓		
	Jakob Thomasius		
	↓		
	Friedrich Leibniz		

Legend for Pedigree of Karl Gegenbaur

Karl Gegenbaur (1826–1903) was born in Würzburg, Germany, and died in Heidelberg, Germany. He emphasized comparative anatomy and extended the idea of homology which suggested an evolutionary model which he favored over the naturphilosophie that looked upon homology from a Platonic or theological interpretation. His text on comparative anatomy was translated into several languages.

Johannes Müller (1801–1858) was born in Koblenz, Germany, and died in Berlin. His father was a shoemaker. He started out with the intention of becoming a priest but was excited about the courses he took in biology and got an MD instead at Bonn. In 1833 he joined the faculty at Humboldt University in Berlin. He was a physiologist and devoted much of his attention to the organ systems involved in the senses especially speech, hearing, and sight.

Rudolph Virchow (1821–1902) was born in Schivelbein, Germany, and died in Berlin, Germany. He got his PhD with Johann Müller. He was a founder of the field of public health and prepared a report on the causes of the high mortality and disease rate of German miners in Silesia. He blamed low wages and supported the 1848 Revolution. He was fired and went to a western principality where he developed the field of pathology, especially using the microscope to study affected tissues. He adopted and modified the cell theory of Remak and argued all cells arise from pre-existing cells and he claimed cancers arose from single cell defects. He called this modification the cell doctrine.

Franz Leydig (1821–1908) was born and died in Rothenberg, Germany. He studied microscopic anatomy and in the testes of animals he found interstitial cells, now named Leydig cells, that are associated with testosterone production during the maturation of spermatozoa.

Asa Gray [1810–1888]

John Torrey [1796–1873]	**Charles Darwin** [1809–1882]	Alphonse de Candolle [1806–1893]
↓		↓
Wright Post [1766–1828]	↓	Augustin Pyramus de Candolle [1778–1841]
↓		↓
John Hunter [1728–1793] ↓	See **Muller** pedigree **Darwin** to **Newton**	Pierre Etienne Vaucher [1763–1814]
William Cheselden [1688–1752]		
↓		
William Cowper [1666–1709]		

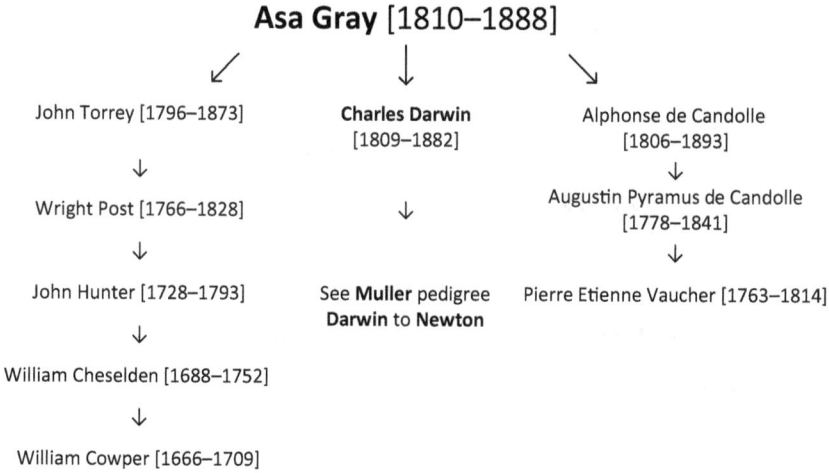

Legend for Pedigree of Asa Gray

Asa Gray (1810–1888) was born in Sauquot, NY, and died in Cambridge, Massachusetts. His father was a tanner and a farmer, neither of which appealed to Asa Gray. He chose medicine and there shifted again to botany because he loved collecting and looking for new varieties and species. He met **John Torrey** who took him on as a student and co-worker. In 1838, at the University of Michigan, he became the first botany professor whose duty was doing full time botany. He made a tour of Europe and purchased 3700 books on botany for the University library. Later he joined the faculty at Harvard. He corresponded widely and numerous letters were with **Charles Darwin**. It was Darwin's 1844 sketch of his theory of evolution by natural selection sent to Gray, that Hooker, and Huxley used to demonstrate Darwin's independent discovery of natural selection when the Linnaean Society published a segment drawn from it and **Wallace's** paper.

John Hunter (1728–1793) was born near Glasgow, Scotland, and died in London, England. He worked mostly in London where his older brother taught him medicine. He was gifted in developing new techniques for studying anatomy. He injected colored fluids into veins of cadavers and demonstrated that the blood supply of the woman and her fetus were separated by a placental membrane. He was one of the first to use artificial insemination by syringe from a husband whose condition (hypospadias) prevented getting his wife pregnant. He performed more than 2000 dissections in his career. He also served in the 7 Years' War and wrote a treatise on the treatment of gunshot wounds.

William Cheselden (1688–1752) was born in Somerby, England, and died in Bath, England. He was an anatomist and had joined "the company of barbers and surgeons" as a student. He supported a name change and in 1745 it became "the company of surgeons" and even later it became "the Royal College of Surgeons." He was successful introducing a surgical technique to remove bladder stones. He also was successful in restoring sight to patients with opaque cataracts. He wrote the first English language anatomy book, in 1712.

William Cowper (1666–1709) was born and died in Peters-field, England. He is best known for describing Cowper's gland (also known as the bulbourethral gland). It lubricates the urethra during arousal and permits a rapid discharge of semen. His career was blemished by his purchase of engraved anatomical plates from book publishers and widows of authors of anatomy books. He then used and altered the plates for his own treatises on anatomy without giving credit to the original artists or anatomists.

Alphonse Pyramus de Candolle (1806–1893) was born in Paris and died in Geneva. His father, also a botanist, encouraged his career. Alphonse de Candolle wrote a book on Geographic Botany (1855) which related plants to their environments, a book very much cited by **Darwin**. He also mentored **Nicolai Vavilov** who extended those studies to centers of origins of domesticated plants.

Augustin Pyramus de Candolle (1778–1841) was born and died in Geneva. He reformed the classification of plants using a functional approach instead of an arbitrary approach (such as petal or anther number in flowers). He coined the term homology for somewhat different structures having a common origin such as petals and leaves. He also described "nature's wars" in which plants replaced other plants in each region. **Darwin** used this as supporting evidence for natural selection.

Jean Pierre Etienne Vaucher (1763–1841) was born and died in Geneva. He was a Protestant pastor who taught botany and later church history. He was especially interested in algae. He demonstrated that some algae have a sexual life that he described as conjugation, with one cell entering another cell to produce a spore or altered cell which could generate a new algal plant.

Ernst Haeckel [1834–1919]

```
      ↙              ↓              ↘
```

| Rudolf Virchow | J. P. Müller | Karl Gegenbaur |
| [1821–1902] | [1801–1858] | [1826–1906] |

```
      ↓           ↙    ↘              ↓
```

| August Froriep | Karl Rudolphi | August Mayer | see pedigree |
| [1849–1917] | [1771–1832] | [1787–1865] | |

```
                   ↓
```

Christian Weigel
[1748–1831]
↓

Johann Erxleben
[1744–1777]
↓

Abraham Kastner
[1719–1800]

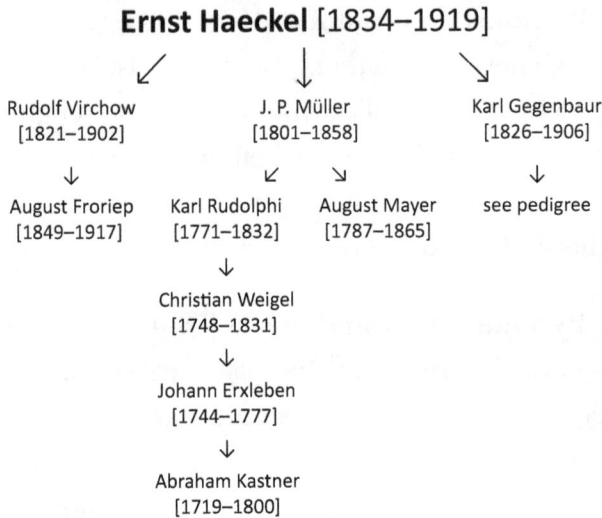

Legend for Pedigree of Ernst Haeckel

Ernst Haeckel (1834–1919) was born in Potsdam and died in Jena, both in Germany. He was a gifted artist whose renderings of calcareous protozoans were much admired. He became a philosopher of biology and championed Darwinian evolution. He coined the phrase "ontogeny recapitulates phylogeny."

Rudolf Virchow (1821–1902) was born in Swirwin, Poland, and died in Berlin, Germany. As a young physician he prepared a report on high morbidity on Silesian miners and blamed it on low wages. It was 1848 and he supported the Socialist revolution. He was fired. He went to western Germany and became a pathologist. He proposed cancer as a cellular disease arising from a single cell origin. He proposed the cello doctrine — all cells arise from pre-existing cells (independently proposed by Remak) — there is no free formation or spontaneous formation of cells. He ran for office and advocated public health programs for all German school children. He proved most Germans were not Aryan in ancestry.

Robert Froriep (1849–1917) was born in Weimar and died in Tübingen, both in Germany. He was an anatomist and an artist, teaching anatomy to art students and illustrating for medical journals. He described rheumatoid arthritis disorders.

Karl Rudolphi (1771–1832) was born in Stockholm, Sweden, and died in Berlin, Germany. He was a founder of the field of helminthology and studied nematode and flatworm morphology. He also demonstrated that cells were individual units and not embedded in a mesh-like communal cell wall.

Christian Ehrenfried Weigel (1748–1831) was born in Stralsund, Sweden, and died in Greifswald, Germany. He developed a refrigeration unit (counter current cooler) for his laboratory work in physiology and taught minerology, botany, and pharmacy.

Johann Christiar Polycarp Erxleben (1744–1777) was born in Quedlinburg, Germany, and died in Göttingen, Germany. He was the founder of the first veterinary medical school in Germany. His mother was the first woman to earn an MD in Germany. He also taught mathematics.

Abraham Kastner (1719–1800) was born in Leipzig and died in Göttingen, both in Germany. He was a mathematician who had written many texts, including a history of mathematics.

Johannes Peter Müller (1801–1858) was born in Coblenz and died in Berlin, both in Germany. He studied the physiology and anatomy of sense organs. And wrote an influential text in physiology.

August Mayer (1787–1865) was born in Gmünd and died in Bonn, both in Germany. He studied comparative anatomy and contrasted the human, chimpanzee, and orangutan anatomies.

Karl Gegenbaur (1826–1903) was born in Würzburg, Germany, and died in Heidelberg, Germany. He stressed the use of homology in his comparative anatomy studies and used this to support the evolutionary history of taxonomic groups.

John B. S. Haldane [1892–1964]

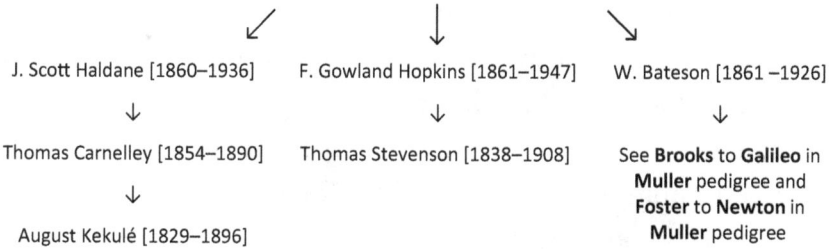

↙	↓	↘
J. Scott Haldane [1860–1936]	F. Gowland Hopkins [1861–1947]	W. Bateson [1861 –1926]
↓	↓	↓
Thomas Carnelley [1854–1890]	Thomas Stevenson [1838–1908]	See **Brooks** to **Galileo** in **Muller** pedigree and **Foster** to **Newton** in **Muller** pedigree
↓		
August Kekulé [1829–1896]		

Legend for J. B. S. Haldane Intellectual Pedigree

John Burdon Sanderson Haldane (1892–1964) was born in Oxford, England, and died in Bhubaneswar, India. He was a geneticist, mathematician, essayist, and social reformer. His book, *Daedalus*, provided the scientific basis for Aldous Huxley's *Brave New World*. He was a Communist until the Lysenko Affair forced him to denounce the USSR for destroying genetics in the Soviet Empire. He first demonstrated linkage in animals and humans. He provided the evidence for balanced polymorphism and the high incidence of sickle cell anemia in malarial areas. With **Sewall Wright** and **R. A. Fisher** he was a founder of the field of population genetics. He was largely self-taught and encouraged to learn on his own by his father. After he moved to India in 1956, he stimulated Indian genetics research with Krishna Dronamraju as his first Indian PhD and his future biographer.

John Scott Haldane (1860–1936) was born in Edinburgh and died in Oxford. He was a physiologist who worked on respiratory gases, often using himself as a guinea pig. He showed how carbon monoxide reacts with hemoglobin to produce the toxic symptoms of CO poisoning. He designed decompression chambers to prevent the bends from nitrogen dissolved under pressure.

Thomas Carnelley (1854–1890) was born in Manchester, England, and died in Aberdeen, Scotland of an abdominal abscess at age 36. He was a chemist mentored in Germany by **Kekulé** at Bonn after his education at King's College in London. He explored the chemical properties to elements in the periodic table and published a book of tables for melting points of elements and compounds.

August Kekulé (1829–1896) was born in Darmstadt, Germany, and died in Bonn, Germany. He became a chemist after hearing **Justus von Liebig's** lectures. His most famous contribution to chemistry was the structure of the benzene ring consisting of six carbon atoms organized as a ring. For philosophers of science he is known for his account of a dream while dozing by a fireplace. He saw a snake eating its tail (an Ouroboros) which suggested the circular structure.

Frederick Gowland Hopkins (1861–1947) was born in Eastbourne, England, and died in Cambridge, England. He got his medical education at Guy's Hospital (now King's College) London and a DSc in chemistry in 1902. He participated in the shift from chemical physiology to biochemistry and was the first Cambridge professor with the title Professor of Biochemistry. He discovered vitamins and recommended adding vitamins A and D to margarine. He showed oxygen is depleted in muscle cells and replaced by lactic acid.

Thomas Stevenson (1838–1908) was born in Rainton, England, and died in London. He was a toxicologist and wrote books on forensic medicine. He cites a "Mr. Steele" as his mentor for inspiring a career in medicine.

William Bateson (1861–1926) was born and died in Cambridge, England. His father was Master of St. John's College. Bateson started his academic career with an interest in embryology and after receiving his MA went to study with **William Keith Brooks** at Johns Hopkins. Brooks suggested he choose heredity as his area of research. Bateson returned and wrote a volume on variation, introducing the idea of homeotic and meristic mutations and their relation to evolution. Bateson championed Mendelism, applied it to both animals and plants and found epistasis (modified Mendelian ratios) and coined the term genetics.

Alfred D. Hershey [1908–1997]

↙   ↓   ↘

Jacques Bronfenbrenner [1883–1953]   Hideyo Noguchi [1876–1928]   **Max Delbrück** [1906–1981]

↓   ↓   ↓

William J. Gies [1872–1956]   Simon Flexner [1863–1946]   See pedigree

↓

Elie Metchnikoff [1845–1916]

↓

Rudolf Leukhart [1822–1898]

↓

Rudolf Wagner [1805–1864]

↓

Johann Schönlein [1793–1864]

↓

Ignaz Döllinger [1770–1841]

↓

to **Galileo** in
Muller pedigree

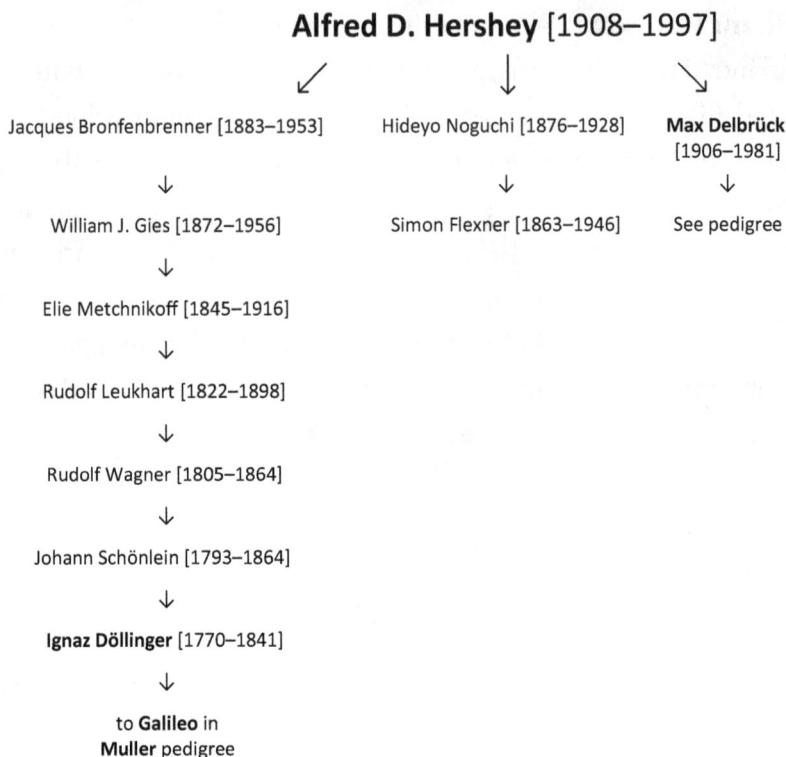

Legend for Intellectual Pedigree for Alfred Hershey

Alfred D. Hershey (1908–1997) was born in Owosso, Michigan, and died in Syosset, New York. He got his BS in chemistry and his PhD in bacteriology at Michigan State University. He worked with **Salvador Luria** at Washington University in St. Louis and he and his colleague, **Martha Chase,** moved to Cold Spring Harbor Laboratory. In 1952 Hershey and Chase published a paper showing viral DNA, not protein, is the basis for generating new phage particles. With **Max Delbrück**, Hershey and **Luria** formed the "phage group" that ushered in a field of molecular biology.

Jacques Bronfenbrenner (1883–1953) was born in Kherson, Ukraine, and died in St. Louis. He studied in Odessa and joined the failed Revolution of 1905 which forced him to flee as a Trotskyite. He worked with **Elie Metchnikoff** at the Pasteur Institute. He moved to the Rockefeller Institute and got his PhD at Columbia with **William Gies**. He also got a Master's degree in public health at Harvard. He worked on vaccines and the immune system, especially phagocytosis.

William John Gies (1872–1956) was born in Reisterstown, Maryland, and died in NY City. He got his PhD at Yale in biochemistry. He studied the composition of saliva. He also played a major role in reforming dental education, changing it from a trade into a profession by promoting adoption of scientific dentistry with criteria like those of medical schools. He established the first biochemistry department in a dental school.

Hideyo Noguchi (1876–1928) was born in Inowashiro, Japan, and died in Accra, Ghana. He got his MD in 1897 with the help of a physician who restored function to his left hand badly burned as a toddler. In 1900 he emigrated to the United States to join **Simon Flexner** at the Rockefeller Institute. He worked on snake venoms and how the immune system responds to them. In 1911 while studying syphilis he isolated the bacterium causing it, *Treponema pallidum*. He used human subjects without permission for studies of a syphilis vaccine with two outcomes. He gave himself (and several of his subjects) syphilis and the outcry led to laws making it a felony to do medical experimentation on humans without informed consent. He left the United States for Ghana and worked on a yellow fever vaccine there until he caught yellow fever and died from it. Several of his colleagues believed he was paranoid.

Elie Metchnikoff (1845–1916) was born in Kharkov, Russia, and died in Paris. His father was in the Imperial Guard and his mother was Jewish. He was encouraged by them and studied at Kharkov before going to Würzburg, Germany to study with **Rudolf Leuckart**. He worked on embryology and found an alternation of haploid and diploid states for nematode worms. He moved to St. Petersburg and got his PhD with **Andrew Kovalesky**. In 1882 he moved to Paris and at the Pasteur Institute in 1883 he discovered phagocytosis of bacteria by white blood cells and this opened the field of immunology. He suffered from depression and twice tried suicide.

Rudolf Wagner (1805–1864) was born in Bayreuth, Germany, and died in Göttingen. His father was a professor. Wagner got his MD at Erlangen and studied with **Johann Schönlein**. His major discovery was the germinal vesicle and its relation to the ovum on the surface of the ovary. His outlook was vitalistic. He embraced **Goethe**'s nature philosophy. He was in ill health most of his life with a degenerative pulmonary disease (possibly alpha-1 antitrypsin dehydrogenase deficiency syndrome).

Rudolf Leuckart (1822–1898) was born in Helmstedt, Germany, and died in Leipzig. He was a parasitologist who studied tapeworms (*Taenia saginata*) and worked out their life cycle in pigs and humans. He also worked out the life cycle of the nematode causing trichinosis. His work led to compulsory meat inspections in Germany and from there to the rest of the world. He showed that Cuvier's "Radiata" was really a mistaken lumping of two phyla — coelenterates and echinoderms.

Alexander Kovalevsky (1840–1901) was born in Varkava, Latvia, and died in St. Petersburg, Russia. He received his doctorate at Heidelberg. He studied tunicates and showed they possessed a notochord and were not mollusks but chordates and possibly ancestral to the vertebrates. He also worked out gastrulation in embryo formation.

Simon Flexner (1863–1928) was born in Louisville, Kentucky, and died in New York City. He got his BA and MD at the University of Louisville and went to Johns Hopkins University to do postdoctoral work in pathology. He was recruited as the founding Director of the Rockefeller Institute in New York. He was a gifted administrator and made the Rockefeller Institute a model of a medical research center with several of its star members winning Nobel prizes. In 1908 he predicted that heart and kidney transplants would be common in human medical practice. His work on using monkeys to study polio virus eventually provided the tools for working out a vaccine.

François Jacob [1920–2013]

↙   ↓   ↘

André Lwoff [1902–1994]	Edouard Chatton [1883–1947]	Jacques Monod [1910–1976]
↓		↓
Otto Meyerhoff [1884–1951]		Georges Tessiers [1900–1972]
↓		↓
Otto Warburg [1883–1970]		Philippe L'Héritier [1906–1994]
↓		↓
Emil Fischer [1852–1919]		**Theodosius Dobzhansky** [1900–1975]
↓		↓
Adolph von Baer [1835–1917]		See pedigree
↓		
Friedrich Kekulé [1829–1896]		

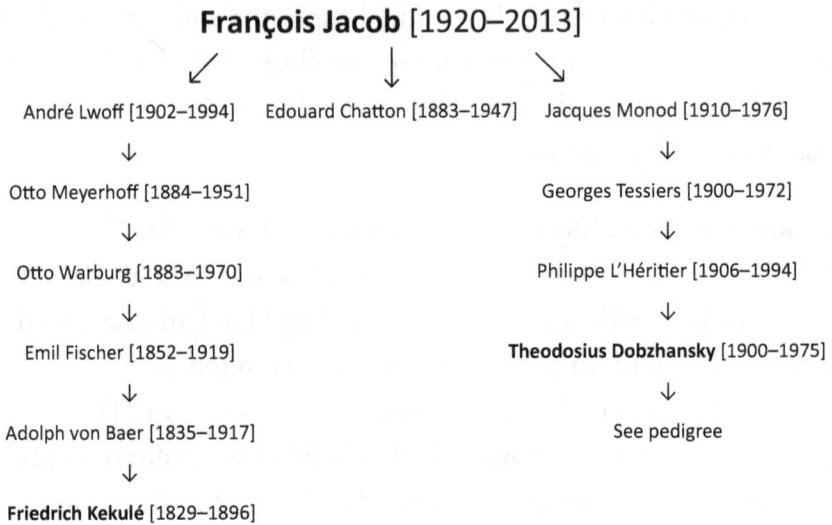

Legend for François Jacob Pedigree

François Jacob (1920–2013) was born in Nancy and died in Paris. His father was a merchant. Jacob was an only child and while his father was observant as a Jew, his mother was secular in her Jewish identification. Their son became an atheist. When WWII broke out, Jacob went with the free French forces to Africa where he was engaged in several battles and severely wounded in Tunisia. After returning to France he continued his medical education getting an MD and then switched to research and got a PhD at the Sorbonne working on bacteria with **Lwoff** and **Monod**. He and Monod worked out the operon model for bacterial regulation of metabolism, switching genes on or off through a molecular process of molecules serving as switches that were activated or repressed by the presence or absence of a substrate such as a sugar molecule.

Edouard Chatton (1883–1947) was born in Romont, Switzerland, and died in Banyuls-sur-mer, France. He was a protozologist who studied the evolution of the mitotic process using ciliates and dinoflagellates. He also contributed to an understanding of nuclear and cytoplasmic relations.

André Lwoff (1902–1994) was born in Ainay-le-Château, France, and died in Paris. He received his MD in 1927 and then got his PhD in 1932 at the Institut Pasteur. He then did postdoctoral work with **Meyerhof** and **David Keinin** at the Rockefeller Institute. He studied lysogeny in bacteria and used ultraviolet to shift the prophage state to the lytic state.

Jacques Monod (1910–1976) was born in Paris and died in Cannes, France. His parents moved to the south of France when he was a toddler. Monod's father was a painter and his mother was an American. He studied with **Teissier, Lwoff, Ephrussi**, and **Rapkine** in Paris. His BS was in 1931 and his PhD was in 1941. He was active in the French underground during World War II. With **François Jacob** he worked out the operon model of gene regulation for metabolism and development. He also wrote *Chance and Necessity* on the role of philosophy in science. His wife was an archaeologist.

Georges Teissier (1900–1972) was born in Paris and died in Roscoff, France. He was one of the few French biologists who was a Darwinian favoring natural selection (most were Lamarckists in their preference). He argued that both microevolution and macroevolution were governed by the same mechanisms and outcomes of natural selection. He favored mathematical approaches to population genetics.

Philippe L'Héritier (1906–1994) was born in Ambert, France, and died in Champétières, France. He received his PhD with **Georges Teissier**. He studied population genetics and designed "cages" in which fruit flies could migrate to different containers.

Using the mutants Bar eyes and ebony body color, he showed these traits favored heterozygotes and maintained polymorphism. He also demonstrated that carbon dioxide sensitivity in fruit flies was due to a virus which he isolated. L'Héritier and Teissier promoted Darwinian evolution and Mendelian genetics in France at a time when Lamarckian biology was widely accepted in France.

Theodosius Dobzhansky (1900–1975) was born in Nemyriv, Ukraine, and died in San Jacinto, California. He was an only child and his father was a mathematician. He received his PhD in 1927 from **Yuri Filipchenko** at St. Petersburg. In 1927 Dobzhansky left Russia for the United States on a Rockefeller fellowship and studied fruit fly genetics with **Morgan** and **Sturtevant**. He followed them to Caltech and worked there to 1940 when he was appointed to Columbia University. His field was population genetics and he wrote, in 1937, *Genetics and the Origin of Species*. It was widely used to promote the new synthesis of classical genetics, cytogenetics, population genetics, and natural selection.

Yuri Filipchenko (1882–1930) was born in Zlyn, Russia, and died in St. Petersburg, Russia. He received his PhD with **Vladimir Shevyakov**. He introduced the terms microevolution and macroevolution. Microevolution involved speciation and macroevolution involved the formation of higher categories

Wilhelm Johannsen [1857–1927]

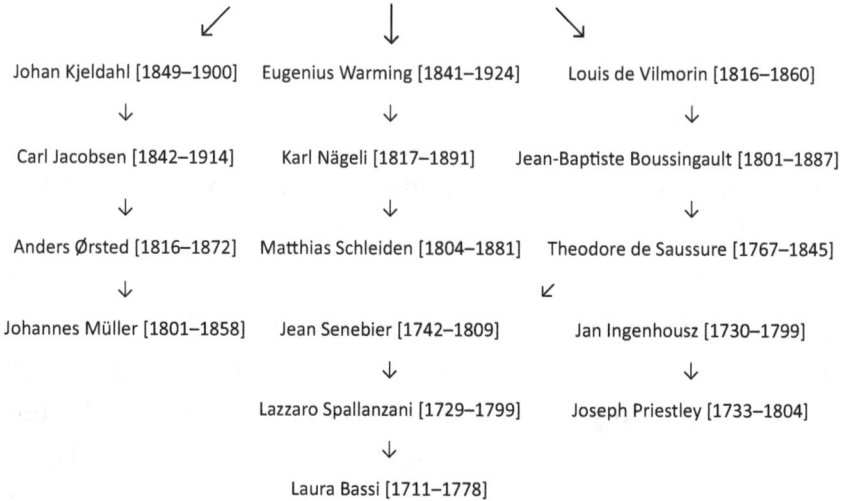

Johan Kjeldahl [1849–1900]	Eugenius Warming [1841–1924]	Louis de Vilmorin [1816–1860]
↓	↓	↓
Carl Jacobsen [1842–1914]	Karl Nägeli [1817–1891]	Jean-Baptiste Boussingault [1801–1887]
↓	↓	↓
Anders Ørsted [1816–1872]	Matthias Schleiden [1804–1881]	Theodore de Saussure [1767–1845]
↓		↙
Johannes Müller [1801–1858]	Jean Senebier [1742–1809]	Jan Ingenhousz [1730–1799]
	↓	↓
	Lazzaro Spallanzani [1729–1799]	Joseph Priestley [1733–1804]
	↓	
	Laura Bassi [1711–1778]	

Legend for Pedigree of Wilhelm Johannsen

Wilhelm Johannsen (1857–1927) was born and died in Copenhagen, Denmark. His father was an army officer. Johannsen studied pharmacology and used ether and chloroform to stimulate germination of seeds. He taught plant physiology at Carlsberg Laboratories. He shifted to the study of inbreeding and selection in beans establishing pure lines and contrasted them to the bell-shaped Gaussian curves of outbred beans. He coined the term genes, genotype, and phenotype. He rejected vitalism but saw the living organism as a product of all its genes but felt it was premature to speculate how an organism is produced by its genes.

Eugenius Warming (1841–1924) was born in Mano, Denmark, and died in Copenhagen, Denmark. He got his PhD in botany at the University of Copenhagen. He is the founder of plant ecology. He studied plant distribution and adaptations in Greenland, Denmark, Venezuela, and the West Indies. He called his field "phytogeography." He identified "plant communities" as descriptive entities in 1895.

Karl Nägeli (1817–1891) was born in Zurich, Switzerland, and died in Munich, Germany. He got his MD at the University of Zurich and studied plant anatomy and physiology, coining the terms phloem, xylem, and meristem. He believed heredity was transmitted by units and corresponded with Mendel (they sent each other seeds of plants with predicted ratios). Nägeli showed that hawkweeds (*Hieracium*) did not give 3:1 ratio as peas did for Mendel. Some 60 years later, hawkweeds were shown to be the rare plant that reproduces by a form of cloning from stimulated, but not fertilized eggs. Mendel, who had a weak ego, abandoned genetics when he found his ratios were not universal.

Matthias Schleiden (1804–1881) was born in Hamburg, Germany, and died in Frankfurt, Germany. His father was a physician. He studied law but a failing practice led to a suicide attempt and shot himself in the head. He survived and switched to medicine with Johannes Müller in Berlin and chose botany as what he liked best. In 1838 he proposed the cell theory and claimed plants were communities of cells. He learned by chance that **Theodor Schwann** found the same in animals and they presented their work jointly. Schleiden popularized botany with books, lectures, and articles.

Jean Senebier (1742–1809) was born and died in Geneva, Switzerland. His father was a merchant. He showed plants produced oxygen (then called dephlogisticated air) and use carbon dioxide (then called carbonic acid). He found that only the green parts of plants produce oxygen.

Lazzaro Spallanzani (1729–1799) was born in Scandiano, Italy, and died in Pavia, Italy. He was ordained a priest and studied spontaneous generation, which he rejected. He showed that boiled water (with or without nutriments added) in sealed glass containers did not undergo fermentation or rotting. He also showed that digestion requires specific juices (gastric juices) and that food was not merely pulverized by grinding and churning into smaller particles.

Laura Bassi (1711–1778) was born and died in Bologna. She was the first female scientist to earn a doctorate degree (1732) and have an academic career. She popularized Newton's work and mentored **Spallanzani**. She married **Giuseppe Verratti** who taught medicine and physics. They had 12 children, only five surviving infancy.

Anders Ørsted (1816–1872) was born in Rudkøbing, Denmark, and died in Copenhagen, Denmark. He came from an illustrious family and his uncle was **Hans Christian Ørsted**, the physicist who worked on electricity and magnetism. He discovered that some plants were parasitic and had two hosts to complete their life cycle. The rust fungus lives one stage on Juniper plants and another stage in pear plants.

Louis de Vilmorin (1816–1860) was born and died in Paris, France. He came from a family of wealthy landowners who took an interest in soil chemistry and biology. Vilmorin used intense selection to develop pure-breeding seeds and was the first to sell seed commercially throughout Europe. He much appreciated the published work of **Jean-Baptiste Boussingault**.

Johannes Peter Müller (1801–1858) was born in Koblenz, Germany, and died in Berlin, Germany. His father was a shoemaker and young Müller thought he would be a priest but shifted to medicine. He received his medical degree in Bonn. He specialized in the physiology of the senses and taught at the University of Berlin.

Jean-Baptiste Boussingault (1801–1887) was one of the founders of soil science. He studied the benefits of crop rotation, especially with legumes as an alternate crop and assigned this to the production of nitrogen compounds brought in from the atmosphere and altered by the cell's metabolism.

Theodore de Saussure (1767–1845) was born and died in Geneva, Switzerland. He demonstrated in 1804 that plants take in carbon dioxide and water and convert it into food with the release of oxygen into the atmosphere.

Jan Ingenhousz (1730–1799) was born in Breda, Netherlands, and died in Calne, England. After receiving his MD, he applied vaccination (using attenuated smallpox from pustules) and applied this successfully to 700 patients. This inspired **Empress Maria Teresa** to have Ingenhousz inoculate her and members of the royal household. His success made him court physician to the Empress. He used his time to repeat **Priestley's** experiments on plants and found that only the green parts of plants produced bubbles when immersed in water and exposed to sunlight. He demonstrated that the bulk of a plant's mass comes from air and not from the minerals in the soil.

Joseph Priestley (1733–1804) was born in Birstall, England, and died in Northumberland, Pennsylvania. He is a founder of English Unitarianism, a theologian, a historian, discoverer of oxygen (independently of Scheele and Lavoisier), inventor of soda water, introduced the time line for the study of history, and used experiments to interpret electricity, the production of oxygen (by focusing the sun's rays on mercuric oxide), and many other works. His home, school, and laboratory were burnt to the ground in Birmingham because of his support for the French Revolution. He was greeted as a hero when he emigrated to the United States but was enfeebled with old age.

Motoo Kimura [1924–1994]

↙   ↓   ↘

Hideki Yukawa [1907–1981]   Hiloshi Kihara [1893–1986]   James Crow [1906–2012]

↓

John T. Patterson [1878–1960]

↓

Charles O. Whitman [1842–1910]

↓

Rudolf Leuckart [1827–1898]

↓

Rudolf Wagner [1805–1864]

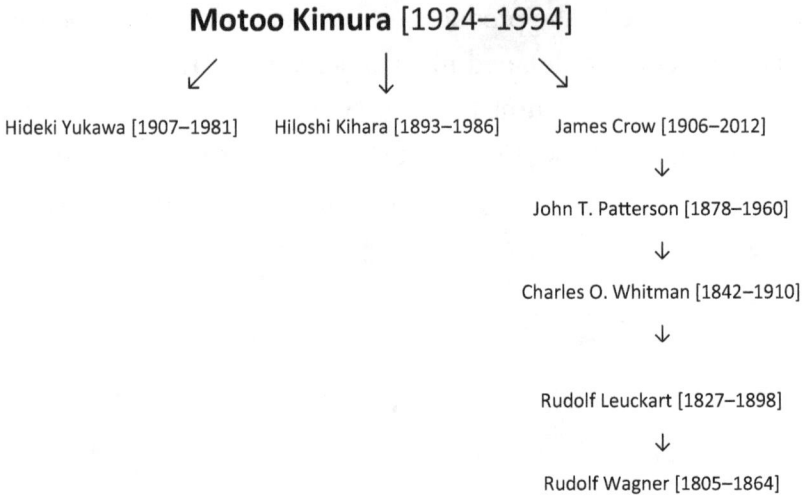

Legend for Pedigree of Motoo Kimura

Motoo Kimura (1924–1994) was born in Okazaki and died in Mishima, both in Japan. His father was a businessman who loved horticulture and encouraged his son to investigate botany. He went to the University of Wisconsin for his graduate work and was mentored by **James Crow**, getting his PhD in 1956. He applied stochastic methods to population genetics. In 1968 he proposed the neutral theory of mutation. He argued most mutations have no detrimental or beneficial effect and are carried along by random drift. He developed amyotrophic lateral sclerosis and died from a fall.

Hideki Yukawa (1907–1981) was born in Tokyo and died in Kyoto, both in Japan. He became a physicist and described the meson for which he won a Nobel Prize. His father was a professor of geology. He was a colleague of **Kimura's** cousin and inspired Kimura to succeed in his field.

Hitoshi Kihara (1893–1986) was born in Tokyo and died in Yokohama, both in Japan. He began college at Hokkaido University in agricultural science and went to Germany to study with **Correns** and came back to focus on wheat. He obtained varieties and showed that the three major varieties were diploid, triploid, and tetraploid sets of chromosomes. He called this common set of genes the genome. He also discovered sex chromosomes in sorrel species and thus some plants, like animals, have sex chromosomes. He is known for his quote: "The history of the earth is recorded in its crust. The history of all organisms is inscribed in the chromosomes."

James Franklin Crow (1906–2012) was born in Phoenixville, Pennsylvania, and died in Madison, Wisconsin. His father taught at Ursinus College. Crow got his PhD at the University of Texas in 1937. **Muller**, **Patterson**, and **Stone** were his mentors. At the University of Wisconsin, he applied mathematics to the concepts of genetic load, aging effect on sperm, inbreeding, and damage to populations from ionizing radiation. He was an atheist.

Joshua Lederberg [1925–2008]

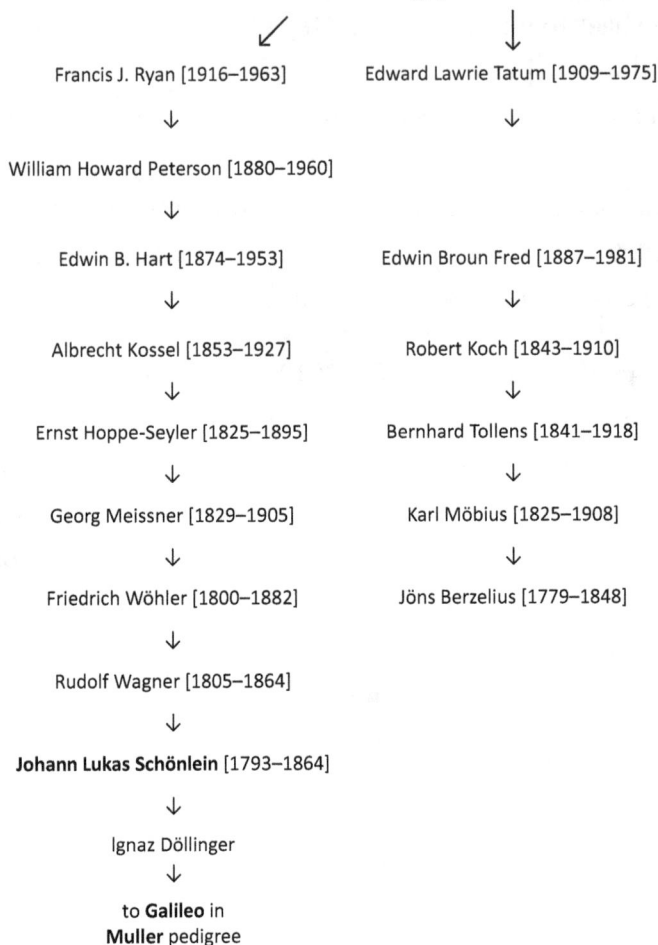

✓   ↓

Francis J. Ryan [1916–1963]   Edward Lawrie Tatum [1909–1975]

↓   ↓

William Howard Peterson [1880–1960]

↓

Edwin B. Hart [1874–1953]   Edwin Broun Fred [1887–1981]

↓   ↓

Albrecht Kossel [1853–1927]   Robert Koch [1843–1910]

↓   ↓

Ernst Hoppe-Seyler [1825–1895]   Bernhard Tollens [1841–1918]

↓   ↓

Georg Meissner [1829–1905]   Karl Möbius [1825–1908]

↓   ↓

Friedrich Wöhler [1800–1882]   Jöns Berzelius [1779–1848]

↓

Rudolf Wagner [1805–1864]

↓

Johann Lukas Schönlein [1793–1864]

↓

Ignaz Döllinger
↓
to **Galileo** in
Muller pedigree

Legend for Joshua Lederberg Pedigree

Joshua Lederberg (1925–2008) was born in Montclair, New Jersey, and died in NY City. His father was a Rabbi and moved the family to New York City. Lederberg attended Columbia University with an intent to earn a MD. During WWII he served in the medical corps and then returned to Columbia and enjoyed research on Neurospora and bacteria. He got his PhD at Yale where he found sexual union in bacteria and was able to map the genes he studied.

He devised techniques (replica plating) to obtain new mutations and he later discovered transduction (the use of virus vectors to carry genes from one bacteria cell or species to another cell). He received a Nobel Prize for his contributions to bacterial genetics. He taught at Wisconsin and became the President of Rockefeller University.

Francis J. Ryan (1916–1963) was born and died in New York City. He attended Columbia University and majored in zoology. He learned of the work of **Beadle** and **Tatum** and went to Stanford to learn Neurospora genetics. He returned to Columbia and studied mutation in bacteria. He travelled around the world introducing microbial genetics to Japan, Israel, and other countries and trained a generation of new microbial geneticists. He was 47 when he died unexpectedly.

William Peterson (1888–1960) No information available.

Edwin B. Hart (1874–1953) was born in Sandusky, Ohio. He was the youngest of 14 children and his mother died when he was an infant and he was raised by his older sisters. He became an agricultural chemist and introduced iodized salt as a means of preventing goiters (and babies born with cretinism). He was mentored for two years by **Kossel**, then at Marburg, Germany.

Albrecht Kossel (1853–1927) was born in Rostock, Germany, and died in Heidelberg, Germany. His father was a Prussian consul. He received his education at the University of Rostock and in Strasburg. He named and isolated the nitrogenous bases of nucleic acid: adenine guanine, cytosine, thymine, and uracil.

Ernst Hoppe-Seyler (1825–1895) was born in Freyburg, Germany, and died in Bodensee, Germany. His father was a bishop. He was orphaned by age 9 and raised by his sister until he was in high school where he was adopted by his teacher whose last name (Seyler) he attached to his birth name (Hoppe). He received his MD training at Halle and Leipzig Universities. He was a founder of medical biochemistry and proved that oxygen binds to hemoglobin and the hemoglobin contains iron. He was also the first to crystallize hemoglobin.

Georg Meissner (1829–1905) was a neuroanatomist with an MD from Göttingen in 1852. He worked on the sensory nerves and found what are called Meissner's corpuscles which provide a sense of touch, pressure, or contact with material objects.

Friedrich Wöhler (1800–1882) was born in Frankfurt, Germany, and died in Göttingen, Germany. He is celebrated for the synthesis of urea, the first organic compound synthesized by chemical means rather than by living cells. Wöhler's organic chemical contributions are often credited with discrediting vitalism, a view that argued living molecules could only be produced by living cells.

Rudolf Wagner (1805–1864) was born in Bayreuth, Germany, and died in Göttingen, Germany. His father was a professor. He named and described the germinal vesicle in the formation of the mammalian egg (the arrested stage of what was later called prophase I of reduction division). Wagner suffered from ill health (tuberculosis) but died from a paralytic stroke.

Legend for second branch of Lederberg's intellectual pedigree

Edward Lawrie Tatum (1909–1975) was born in Boulder, Colorado, and died in New York City. He attended the University of

Chicago for two years and then transferred to the University of Wisconsin for his BA (1931) and PhD (1934). In 1937 he went to Stanford University to study with **Beadle**. In 1945 he mentored **Lederberg** at Yale. He provided the biochemical background for the work Beadle had initiated in fruit flies. The shift to Neurospora (with Beadle) and bacteria (with Lederberg) led to a field of biochemical genetics and the construction of biochemical pathways.

Edwin B. Fred (1887–1981) was born in Middleburg, Virginia, and died in Madison, Wisconsin. His grandparents lost their estate after the Civil War and moved to Oklahoma. He was a mischievous child and got expelled from school. He did his undergraduate work at Virginia Polytechnic School and went to Göttingen for his PhD. He liked bacteriology and studied with **Robert Koch** and **Bernhard Tollens**. When he returned to the United States, he devoted his research to soil bacteria and the fixation of nitrogen in plant roots.

Robert Koch (1843–1910) was born in Hanover, Germany, and died in Baden-Baden, Germany. He received his MD from Göttingen University with a thesis on succinic acid. He served as a surgeon during the Franco-Prussian War and on his return studied bacterial infections, beginning with anthrax. He worked out Koch's postulates on proving a specific bacterial species produces a specific disease. He also used agar as a culture medium and showed how pure cultures could be isolated for study.

Bernhard Tollens (1841–1918) was born in Hamburg, Germany, and died in Göttingen, Germany. He showed that carbohydrates were composed of sugar molecules and characterized the type of sugar in different types of carbohydrates.

Karl Möbius (1825–1908) was born in Eilenberg, Germany, and died in Berlin. He got his doctorate at Kiel University. He studied the feasibility of culturing oysters in Germany and produced a monograph on the ecology of oysters and introduced the term ecosystem (coenoses) to describe his findings and why it would not work in Germany.

Jöns Berzelius (1779–1848) was born in Östergötland, Sweden, and died in Stockholm, Sweden. He was educated in Uppsala University. He became a chemist and found the elements silicon, thorium, selenium, and cerium. He introduced the concept of atomic weights and proposed the notation presently used of a 1 or 2 letter abbreviation for an element with a subscript for the number of atoms involved in a compound. Thus, CO_2 represents carbon dioxide.

Edward B. Lewis [1918–2004]

    ↙   ↓

Clarence Paul Oliver [1898–1991]   Alfred Henry Sturtevant [1891–1970]

    ↓   ↓

Hermann Joseph Muller [1890–1967] → Thomas Hunt Morgan [1866–1945]

                      ↓

See **Muller** pedigree from **Morgan** to **Galileo**

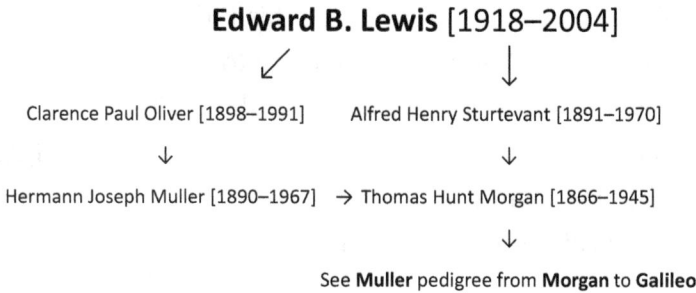

Legend for Pedigree of E. B. Lewis

Edward B. Lewis (1918–2004) was born in Wilkes-Barre, Pennsylvania, and died in Pasadena, California. The B. is on his birth certificate but was not entered for his father's full middle name which was Butts. His father was a watch-maker. He attended the University of Minnesota for his undergraduate degree and then served in the US Air Force preparing meteorological reports. He obtained his PhD with **Sturtevant** at Caltech in 1942. Lewis studied pseudoallelism and his most significant gene complex was the bithorax series. He related the different alleles to mutations at specific segments of the gene nest and showed they involved embryonic regulators of the abdomen, wing, and leg development. His was the first definite proof of embryonic regulation by genes in fruit flies. He received a Nobel Prize for that work. He also studied Hiroshima data and showed the effects of radiation to leukemia were linear.

Clarence Paul Oliver (1898–1991) was born in Dexter, Missouri, and died in Austin, Texas. In WWI he was drafted and became a sergeant and was captured by the Germans about a month before the Armistice was declared. He studied history at the University of Texas intending to go to law school and supported himself by washing fruit fly bottles in **Muller's** laboratory at Austin, Texas. He did his dissertation research on the relation of mutations induced to dosage of X-rays administered and got his PhD in 1931. He taught at the University of Minnesota where he found instances of pseudoallelism (lozenge eyes especially). He shifted his interest, again, to human genetics, and served as Director of the Dwight Institute of Genetics. He retired in 1971 and spent his last research and teaching years at the University of Texas in Austin.

Alfred Henry Sturtevant (1891–1970) was born in Jacksonville, Illinois, and died in Pasadena, California. When he was a boy his family moved to Alabama where his father became a farmer. His grandfather was a College President and Sturtevant's brother was a professor at Columbia University. Sturtevant stayed at his brother's apartment in NYC and through him met Morgan who encouraged his interest in genetics. He asked **Morgan** if he could use his data to work out their unusual ratios. He returned with a map of the six genes that Morgan had studied, all of which were on the X chromosome and Sturtevant (then aged 18) had mapped. It became the basis for his PhD with Morgan. Later Sturtevant studied *D. simulans* and compared its map to *D. melanogaster* and showed most of the gene sequences were intact, but he identified several dozen breakages where past chromosome rearrangements had occurred.

Salvador Luria [1912–1991]

↙ ↓ ↘

Enrico Fermi [1901–1954]   Max Delbrück [1906–1981]   Giuseppe Levi [1872–1965]

↓   ↙   ↓   ↓

Max Born [1882–1970]   Niels Bohr [1885–1962]   Alessandra Lustig [1857–1937]

↓   ↓

Carl Runge [1856–1927]   Christian Christianson [1843–1912]

↓   ↓

Karl Weierstrass [1815–1897]   Carl V. Holten [1818–1886]

↓   ↓

Christoph Gudermann [1798–1852]   Hans Ørsted [1777–1851]

↓

Carl Gauss [1777–1855]

Legend for Academic Pedigree of Salvador Luria

Salvador Luria (1912–1991) was born in Turin, Italy, and died in Lexington, Massachusetts. He was a Sephardic Jew and studied medicine in Turin and was mentored by **Giuseppe Levi**. He went to Rome to work with **Fermi** but had to leave when Mussolini denied academic status to Jews. He went to Paris, but the Germans invaded, and he escaped from Marseilles to the United States. He met **Hershey** and **Delbrück** at Cold Spring Harbor and worked with Delbrück at Vanderbilt. In 1943 he was hired by Indiana University and worked there until 1950 when he moved to Illinois. He finished his career at MIT. He worked out the life cycle of bacteriophage T4 and showed mutations to resistance arose before exposure to such agents. He discovered restriction enzymes and agents called bacteriocins that produced holes in the cell membrane of bacteria.

Enrico Fermi (1901–1954) was born in Rome, Italy, and died in Chicago, Illinois. He received his PhD with **Max Born** and used neutron bombardment to produce trans-uranium elements. He found instead nuclear fission that he realized could be regulated to serve as a nuclear reactor to generate electricity. He also studied proprieties of the neutrinos. He was a key contributor in the Manhattan Project to produce an atomic bomb, but he was a critic of the effort to develop a hydrogen bomb. His most noted students are **C. N. Yang**, **T. D. Lee**, and **E. Segre**.

Max Born (1882–1970) was born in Breslau, Germany, and died in Göttingen, Germany. He received a PhD in solid state physics studying elastic bodies. He contributed to quantum theory and he made Göttingen a world center for physics. His students include **Delbrück, Jordan, Goeppert-Mayer, Oppenheimer,** and **Weisskopf**. He also mentored **Fermi, Heisenberg, Pauli, Teller,** and **Wigner**. After Hitler came to power he went to Scotland and taught physics at the University of Edinburgh. After his retirement in 1952 he returned to Germany.

Carl Runge (1856–1927) was born in Bremen, Germany, and died in Göttingen, Germany. His father was the Danish consul to Havana, Cuba. In 1880 Runge got his PhD in mathematics in Berlin. He worked on vector analysis. He later worked on spectral lines of elements with Karl **Weierstrass**.

Karl Weierstrass (1815–1897) was born in Steinfeld, Germany, and died in Berlin, Germany. He was a gifted mathematician who left college after one year and studied modern analysis. He also took courses with **Christoph Gudermann**. He taught in high school (gymnasium) and introduced spherical geometry to his students. His work on elliptic functions gained him an honorary PhD in mathematics.

Carl Friedrich Gauss (1777–1855) was born in Braunschweig, Germany, and died in Göttingen, Germany. Gauss was a child prodigy whose mother was illiterate and his father working class. He taught himself mathematics and contributed to number theory and statistics. He did not like to teach. His most famous mathematical contribution is the bell curve used in statistics and studies of distribution of traits in genetics and behavioral studies.

Edward Laurens Mark [1847–1946]
↓

Rudolf Leuckart [1822–1898]

↓

Rudolf Wagner [1805–1864]

↙   ↘

Johann Schönlein [1793–1864]   Georges Cuvier [1769–1832]

↓   ↙

Ignaz Döllinger [1770–1841]

↓

See **Muller** pedigree from **Cuvier** to **Galileo**

Legend for Intellectual Pedigree of Edward Laurens Mark

Edward Laurens Mark (1847–1946) was born in Hamlet, New York, and died in Cambridge, Massachusetts. He obtained a BSc degree at the University of Michigan and his PhD at Leipzig. He studied cytology and cell biology and wrote a 400-page monograph on the gastropod Limax. He had many notable students get PhDs under his mentoring, including **Davenport**, **Castle**, and **Jennings**.

Rudolf Leuckart (1822–1898) was born in Helmstedt, Germany, and died in Leipzig, Germany. He made parasitology a science, working out hosts and life cycles for tapeworms, liver flukes, and nematodes.

Johann Schönlein (1793–1864) was born and died in Bamberg, Germany. He got his MD studying in Jena, Göttingen, and Würzburg. He was an early investigator of fungal diseases and identified ringworm as an infection of *Trichophyton Schönleinii*. He gave the name tuberculosis to the disease which was then called consumption. He introduced the study of mycology (fungi) to his medical classes.

Barbara McClintock [1902–1992]

↙   ↓   ↘

Claude B. Hutchison [1885–1980]

Rollins Adams Emerson [1873–1947]	Lester Sharp [1887–1961]
↓	↓
Edward Murray East [1879–1938]	Victor Gregoire [1870–1938]
↓	↓
Cyril George Hopkins [1866–1919]	Jean Baptiste Carnoy [1836–1899]
↓	↓
George Chapman Caldwell [1834–1907]	Carl Zeiss [1816–1888]
↓	↓
Friedrich Wöhler [1800–1882]	Friedrich Körner [1778–1847]
↓	

Jöns Jackob Berzelius [1779–1848]

↓

Johann Afzelius [1753–1837]

↓

Törborn Olaf Bergman [1735–1784]

↓

Carl Linnaeus [1707–1778]

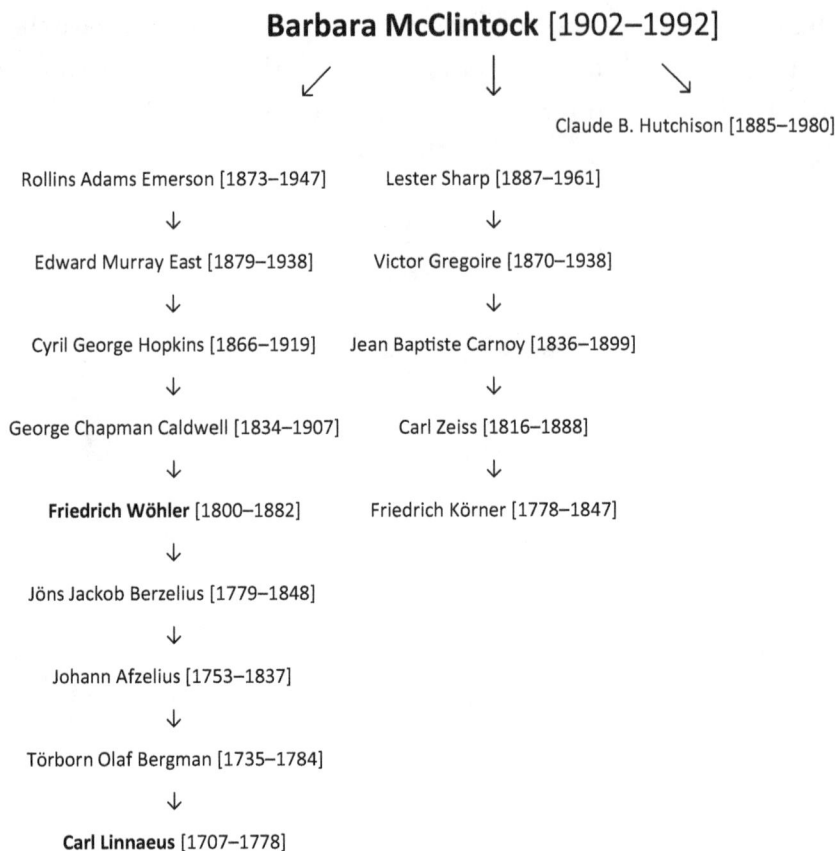

Legend for Intellectual Pedigree for Barbara McClintock

Barbara McClintock (1902–1992) was born in Hartford, Connecticut, and died in Huntington, New York. Her father was a homeopathic physician and she grew up in Brooklyn, attending Erasmus Hall High School. Over her mother's objections she attended college, choosing Cornell University. She chose botany for her major and received her BSc in 1924. Her undergraduate mentor was **Claude B. Hutchison** and her graduate sponsor, also at Cornell, was **Rollins Emerson**. She obtained her cytological skills from **Lester Sharp**. McClintock had difficulty finding an academic position. She worked for many years at the University of Missouri but despite an impressive number of publications she could not get tenured. In 1950 she moved to Cold Spring Harbor Laboratory and spent the rest of her career there. McClintock worked out the linkage maps for maize (*Zea mays*). She also demonstrated that crossing over was associated with chiasma formation and a recombination of visibly distinct parts of two homologous chromosomes. She worked out the cytogenetics of ring chromosomes and the fate of broken chromosomes when they form dicentric chromosomes (the breakage-fusion-bridge cycle). Her Nobel Prize winning work was on genes that regulated other genes by moving from one chromosome to another not by meiotic recombination but by detachment and insertion ("jumping genes"). These were confirmed when microbial geneticists detected them. Her "controlling elements" became transposons. She received a Nobel Prize for contributions to cytogenetics.

McClintock branch 1

Claude Burton Hutchison (1885–1980) was born in Chillicothe, Missouri, and died in Berkeley, California. He grew up on his father's farm. He received his MS in agricultural science from the University of Missouri. His only doctoral degree was in law (1937). He taught at the University of Missouri and at Cornell before moving to the University of California at Davis. He shifted to administration and became Dean of the College of Agriculture at Davis and later went on to become Mayor of Berkeley, California. While he was active in research, he studied aleurone and endosperm traits in maize, including a paper with **Emerson** in 1921.

McClintock branch 2

Rollins Adams Emerson (1873–1947) was born in Pillar Point, NY, and died in Ithaca, NY. He did his undergraduate work in the University of Nebraska and went to Harvard University for his PhD (1907) with **Edward Murray East**. Emerson studied gene mutations associated with color, especially of corn kernels. Emerson was head of the corn genetic group at Cornell which included **McClintock, George Wells Beadle, Marcus Rhodes, and Harriet Creighton**. Like McClintock, Beadle went on to win a Nobel Prize.

Edward Murray East (1879–1938) was born in Du Quoin, Illinois. His father was a mechanical engineer. He attended the University of Illinois getting his BA and a PhD (1913) with **Cyril George Hopkins** in soil chemistry. He worked at the Agricultural Field Station in Connecticut where he studied tobacco, tomato, and corn genetics. He developed a commercially successful way to produce hybrid corn after reading Shull's papers on hybrid corn. He was recruited by Harvard University and made inbreeding and outbreeding the focus of his research career. East was also interested in eugenics and wrote books and articles promoting a racist and bigoted branch of that field. His work was influential in Congress passing the restrictive immigration acts of 1921 and 1924 which were based on racial and ethnic bias.

Cyril George Hopkins (1866–1919) was born in Chatfield, Minnesota, and died overseas in Gibraltar. He was raised in S. Dakota and studied agricultural chemistry at the University of S. Dakota. He got his PhD at Cornell with **George Chapman Caldwell** and then spent a year in Göttingen, Germany with chemist **Bernhard Tollens**.

Branch from Caldwell

George Chapman Caldwell (1834–1907) was born in Framingham, Massachusetts, and died in Ithaca, New York. He studied botany at Harvard University and went to Heidelberg Germany, spending a year (1885) with chemist **Robert Bunsen**. He then shifted to Göttingen to get a PhD in the laboratory of **Friedrich Wöhler** (1857). Caldwell taught at Cornell University, Antioch College, and Pennsylvania State University.

Friedrich Wöhler (1800–1882) was born in Hesse-Kassel and died in Göttingen, Germany. He received his MD at Heidelberg University and worked with **Leopold Gmelin** who recommended him to go to Stockholm to work with **Jöns Jacob Berzelius**. Wöhler used his chemical knowledge to set up a chemistry laboratory at Göttingen. His most famous work was synthesizing urea from sodium cyanate. This was the first demonstration that organic chemicals could be synthesized from non-living molecules. It helped launch the field of organic chemistry.

Jöns Jacob Berzelius (1779–1848) was born in Vaversunda, Sweden, and died in Stockholm. He received his MD at Uppsala University. He was the first to work out atomic weights of elements. He also discovered the elements silicon, selenium, thorium, and cerium. He introduced the chemical notation for elements and compounds used today.

Johan Afzelius (1753–1837) was born in Larv, Sweden, and died in Uppsala. He studied and collected minerals, going on expeditions to Norway, Denmark, and Russia. At Uppsala he studied formic acid (which he identified as the stinging agent from ants), nickel, and oxalic acid.

Törborn Olaf Bergman (1735–1784) was born in Katrineberg, Sweden, and died in Medevi, Sweden. He studied quantitative reactions in chemicals. He discovered bismuth. He worked with **Carl William Scheele** and **Carl Linnaeus** on numerous joint projects.

Carl William Scheele (1742–1786) was born in Stralsund, Pomerania, and died in Köping, Sweden. He was a pharmacist who studied the composition of medicinal products. He discovered barium, magnesium, and tungsten. He independently discovered oxygen (Priestley and Lavoisier are also credited). He was a champion of the phlogiston theory of combustion that was rejected by Lavoisier.

Carl Linnaeus (1707–1778) was born in Rashult, Sweden and died in Uppsala, Sweden. He is best known for his system of classification using Latin genus and species names and devising higher categories of taxonomy using shared morphology.

McClintock's branch from Lester Sharp

Lester Whyland Sharp (1887–1961) was born in Saratoga Springs, New York, and died in Nuevo, California. His family had moved to Michigan and Sharp attended Alma College for his BS. He entered the University of Chicago and obtained a PhD in 1912 in plant morphology. He took a postdoctoral trip to the University of Louvain in Belgium where he was mentored by **Victor Grégoire**. Sharp's major contribution was the first text in cytology and his gift for teaching cytology. He supervised McClintock's first cytogenetic work.

Victor Grégoire (1870–1938) was a priest and botanist at Louvain University. He worked with fungi anatomy. Sharp learned microscopic techniques from his laboratory. Grégoire learned his techniques from **Jean Baptiste Carnoy**.

Jean Baptiste Carnoy (1836–1899) was born in Rumillies, Belgium, and died in Switzerland. He was a scientific instrument maker and studied microscope design with Carl Zeiss.

Carl Zeiss (1816–1888) was born in Weimar, Germany, and died in Jena. He learned instrument making from his father who mostly made art objects from exotic wood. He was sent to specialty schools by his father where he was attracted to the development of microscopes. In 1846 he moved to Jena and he began making microscopes and selling them to university professors. As his work became appreciated, sales mushroomed, and Zeiss became the most widely used source for academic microscope sales and in introducing improvements in their versatility for research.

Friedrich Körner (1778–1847) made scientific instruments, including microscopes. It was in his workshop that Carl Zeiss took an interest in microscope design and efficiency.

Victor McKusick [1921–2008]
↓　　　　　↓

Bentley Glass [1906–2005]　　Harold Jeghers [1904–1990]

↓

H. J. Muller [1890–1967]

↓

See **Muller** pedigree

Legend for Intellectual Pedigree of Victor McKusick

Victor A. McKusick (1921–2008) was born Parkman Maine; he died in Towson, Maryland. His father was a dairy farmer and his mother a teacher. His identical twin brother became a Supreme Court Justice in Maine. McKusick went to Bard College but transferred to Johns Hopkins in his senior year as an early admit candidate to medical school. He specialized in cardiology. He worked with the Amish on genetic disorders (they have a high inbreeding population) and developed human genetics at Johns Hopkins consulting with **Bentley Glass**. He organized a summer course at Bar Harbor in Maine which introduced medical genetics to participants. He helped organize the human genome project and he edited a standard reference work, *Mendelian Inheritance in Man* (now online), that was used by genetic counselors and pediatricians.

Bentley Glass (1906–2005) was born in China and died in Colorado. His parents were missionaries. He got his BA at Baylor and his PhD with **H. J. Muller** at the University of Texas. He taught at Johns Hopkins, Stephens College, and Gaucher College before becoming Vice-President of Stony Brook University in New York. He studied fruit fly genetics and human genetics, calculating gene frequencies in population (30% of the genes in African Americans are derived from mating with whites over the past three centuries).

He served as presidents of Phi Beta Kappa, AAAS, and the AAUP. He was editor of the *Quarterly Review of Biology*.

Hermann Joseph Muller (1890–1967) was born in New York City, NY, and died in Indianapolis, Indiana. (See Muller pedigree for full details.)

Harold Jeghers (1904–1990) was born in Jersey City, New Jersey, and died in North Easton, Massachusetts. He got his MD at Western Reserve University and taught medicine at Boston College, Georgetown, and Tufts. He described independently, what is called Peutz-Jeghers syndrome. It is an autosomal dominant mutation causing polyposis of the intestines and pigmented spots on the lips and mucus membranes.

Gregor Mendel [1822–1884]

↙	↓	↘
Christian Doppler [1803–1853]	Franz Unger [1800–1870]	Andreas von Ettingshausen [1796–1878]
↓	↓	↓
Simon von Stampfer [1792–1864]	Lorenz von Vest [1776–1840]	Ignaz Lindner [1777–1835]
		↓
		Juri von Vega [1754–1802]

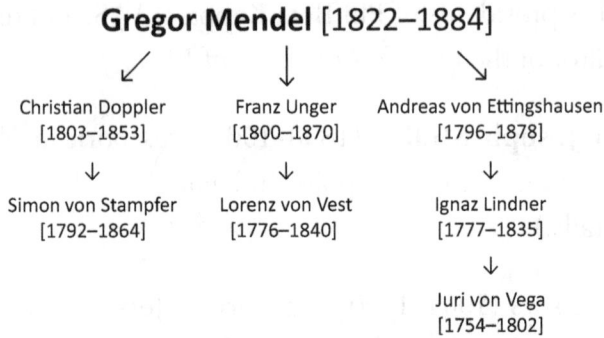

Legend for Gregor Mendel Pedigree

Gregor Mendel (1822–1884) was born in Heinzendorf, Czechoslovakia, and died in Brno, Czechoslovakia. His father was a farmer. Mendel had a gift for learning science, but lack of funds made him choose teaching as an Augustinian monk to live a life in science. He was sent to the University of Vienna where he took courses in the natural sciences, including those of **Christian Doppler** (physics), **Franz Unger** (botany) and **Andreas von Ettingshausen** (mathematics). On his return he began a series of experiments over 9 years leading to a publication on hybridization (1865) which contain his famous distribution laws of segregation and independent assortment based on inferred hereditary particles in the cell.

Christian Doppler (1803–1853) was born in Salzburg, Austria, and died in Venice, Italy. His father was a stonemason, a profession young Doppler could not enter because of frail health (a pulmonary condition perhaps cystic fibrosis). He had skills in mathematics and attended Prague Polytechnic Institute and taught mathematics and physics at several schools. When he was 38, he published a paper on the relation of wavelength of light and distance travelled (the redshift or Doppler Effect). His last years doing research were at the University of Vienna. He went to Venice for his health but died there.

Simon von Stampfer (1792–1864) was born in Matrei, Austria, and died in Vienna. His father was a weaver. Stampfer preferred to study mathematics and he worked out equations to predict eclipses. He also invented in 1832, the "stroboscope" a forerunner of motion pictures using discs with slits and painted or photographed images that when rotated and looking through the slits created an illusion of objects in motion.

Franz Unger (1800–1870) was born in Leutschach, Austria, and died in Graz, Austria, where he was a professor of Botany. He had obtained an MD at Vienna and devoted his research to plant paleontology and plant physiology. In 1851 he proposed that a plant evolution had occurred based on the fossil plants present in different rock strata. He also argued that the traits of plants were probably due to invisible units within the cells of plants, an idea he taught when Mendel took his courses.

Lorenz von Vest (1776–1840) was born in Klagenfurt, Austria, and died in Graz, Austria. He got his MD at Freiburg and Vienna. In 1811 he became the first science professor at Graz. It took 15 years before he had a graduate student wishing to make botany a lifetime career. He taught chemistry and botany. He studied minerals present in meteoroids.

Andreas von Ettingshausen (1796–1878) was born in Heidelberg, Germany, and died in Vienna, Austria. He taught mathematics and physics. He was the first to design a machine run on electricity. He also contributed to mathematics studying combinatorial analysis, a field that **Mendel** used after taking his course. **Ernst Mach** was also a student of von Ettingshausen.

Ignaz Lindner (1777–1835) taught at the Imperial Royal Technical Military Academy in Austria.

Juri von Vega (1754–1802) was born in Slovenia and died near Vienna in an accident. He served in the artillery and taught at the Royal Technical Military Academy. His mathematical education was at Ljubljana. He wrote several books on logarithms and trigonometric functions.

Jacques Monod [1910–1976]
↓

Boris Ephrussi [1901–1979]

↙   ↘

Robert Chambers [1881–1957]   George Beadle [1903–1989]

↓   ↓

Richard Hertwig [1850–1937]   Franklin D. Keim [1885–1956]

↓

Ernst Haeckel [1834–1919]

↓

Albert Kölliker [1817–1905]

↓

Johannes P. Müller [1861–1858]

↓

August F. J. K. Mayer [1787–1865]

Legend for Jacques Monod Pedigree

Jacques Monod (1910–1976) was born in Paris, France, and died in Cannes, France. His father was a French Huguenot and his mother was American. At an early age, Monod sought and found many mentors for his career including **Georges Teissier, André Lwoff, Louis Rapkine, Boris Ephrussi**, and **Alfred Sturtevant**. He got his PhD at the Sorbonne and spent a year, 1934, at Caltech with **Morgan** and his students. Monod spent WWII in the French underground. Monod also enjoyed philosophy and explored "chance and necessity" in the life sciences. His work with **Francois Jacob** led to the operon model of gene regulation.

Boris Ephrussi (1901–1979) was born in Moscow, Russia, and died in Gif-sur-Yvette in France. His family emigrated to

France after the Russian Revolution. His father was a chemical engineer. Ephrussi studied sea urchin eggs and their embryology using micromanipulation of cells. He went to Caltech and demonstrated to **Beadle** how to use fruit fly larval eye buds for transplantation studies that Beadle applied to eye color. After WWII Ephrussi shifted to yeast genetics and nucleocytoplasmic relations.

Robert Chambers (1881–1957) was raised by Canadian missionaries who were in Erzurum, Turkey, where he was born. He died in Concord, New Hampshire. On their return to Canada, he attended Queen's University in Ontario. For his PhD he went to Munich, Germany and studied with **Richard Hertwig**. In 1921 he invented a micromanipulator that he used to carry out microdissection of cells and organelles. He taught at Cornell Medical School and NYU. He was also active each summer at Woods Hole Marine Biology Laboratory.

Richard Hertwig (1850–1937) was born in Friedelberg and died in Schlederloh, both in Germany. He received his doctorate at Bonn. His older brother, **Oscar Hertwig**, was also a prominent biologist and they collaborated on some of their studies. Richard Hertwig was the first to describe and demonstrate that a single sperm entered a single egg during fertilization in the sea urchin. He called the resulting product, a zygote. Richard Hertwig also developed a theory of organogenesis in relation to the embryonic coelom.

Ernst Haeckel (1834–1919) was born in Potsdam and died in Jena, both in Germany. He studied medicine at Würzburg and Berlin. He was an accomplished artist and used his skills to classify coelenterate (Cnidarian) anatomy and complex protozoan organelles. He also became an advocate for **Darwin's** evolutionary views and introduced the "biogenetic" law that "phylogeny recapitulates ontogeny" or in other words, the sequence of events in the evolutionary emergence of organisms is like the sequence of events from fertilization to adult organisms.

Albert von Kölliker (1817–1905) was born in Zurich, Switzerland, and died in Würzburg, Germany. He received his MD in Heidelberg. He devoted his career in microscopy and helped create the field of histology (the cells that form organ tissues) classifying smooth, cardiac, and voluntary (striated) muscles. He demonstrated that axons and dendrites were extensions of individual nerve cells.

Johannes Müller (1801–1858) was born in Koblenz, Germany, and died in Berlin, Germany. His father was a shoemaker. He started out with the intention of becoming a priest but was excited about the courses he took in biology and got an MD instead at Bonn. In 1833 he joined the faculty at Humboldt University in Berlin. He was a physiologist and devoted much of his attention to the organ systems involved in the senses especially speech, hearing, and sight.

August F. J. K. Mayer (1787–1865) was born in Gmünd, Germany, and died in Bonn, Germany. He got his MD in 1812 at Tübingen and was a professor at Bonn. He studied the composition of blood and contributed to the field of comparative anatomy. He showed the anatomical relations among humans, orangutans, and chimpanzees. When the Neanderthal man was first discovered in Germany in 1856, he analyzed the bones and concluded that it was an earlier species of humans.

George Beadle (1903–1989) was born in Wahoo, Nebraska, and died in Pomona, California. His parents were farmers and Beadle thought he would be a better farmer if he went to agricultural school for his training. His teacher, **Franklin Keim**, recognized his talent and recommended Beadle for graduate work at Cornell with **R. A. Emerson**. There, Beadle flourished, studying maize genetics and cytology. He went to Caltech to work with Morgan's group, especially **Sturtevant**, and met **Boris Ephrussi** who was visiting from France. With Ephrussi, Beadle began his studies of biochemical genetics, first in eye color formation and then (with **E. L. Tatum**) using Neurospora. This led to the "one gene: one enzyme" model of gene function and the use of microbes to work out biochemical pathways.

Franklin Keim (1885–1956) was born in Hardy, Nebraska. He received his MS degree at the University of Nebraska and his PhD at Cornell. He studied hybrid wheat and its formation.

Thomas H. Morgan [1866–1945]

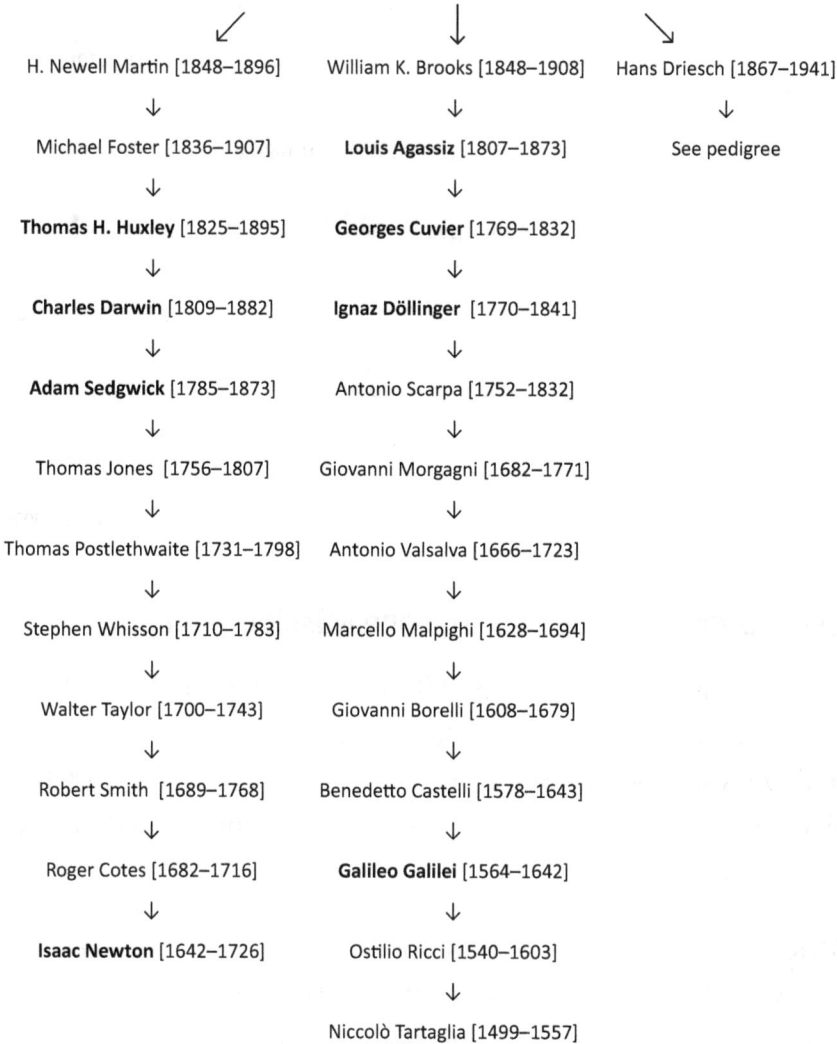

↙	↓	↘
H. Newell Martin [1848–1896]	William K. Brooks [1848–1908]	Hans Driesch [1867–1941]
↓	↓	↓
Michael Foster [1836–1907]	**Louis Agassiz** [1807–1873]	See pedigree
↓	↓	
Thomas H. Huxley [1825–1895]	**Georges Cuvier** [1769–1832]	
↓	↓	
Charles Darwin [1809–1882]	**Ignaz Döllinger** [1770–1841]	
↓	↓	
Adam Sedgwick [1785–1873]	Antonio Scarpa [1752–1832]	
↓	↓	
Thomas Jones [1756–1807]	Giovanni Morgagni [1682–1771]	
↓	↓	
Thomas Postlethwaite [1731–1798]	Antonio Valsalva [1666–1723]	
↓	↓	
Stephen Whisson [1710–1783]	Marcello Malpighi [1628–1694]	
↓	↓	
Walter Taylor [1700–1743]	Giovanni Borelli [1608–1679]	
↓	↓	
Robert Smith [1689–1768]	Benedetto Castelli [1578–1643]	
↓	↓	
Roger Cotes [1682–1716]	**Galileo Galilei** [1564–1642]	
↓	↓	
Isaac Newton [1642–1726]	Ostilio Ricci [1540–1603]	
	↓	
	Niccolò Tartaglia [1499–1557]	

Legend for Thomas Hunt Morgan

See H. J. Muller reference pedigree.

Christiane Nüsslein-Volhard [b. 1942]

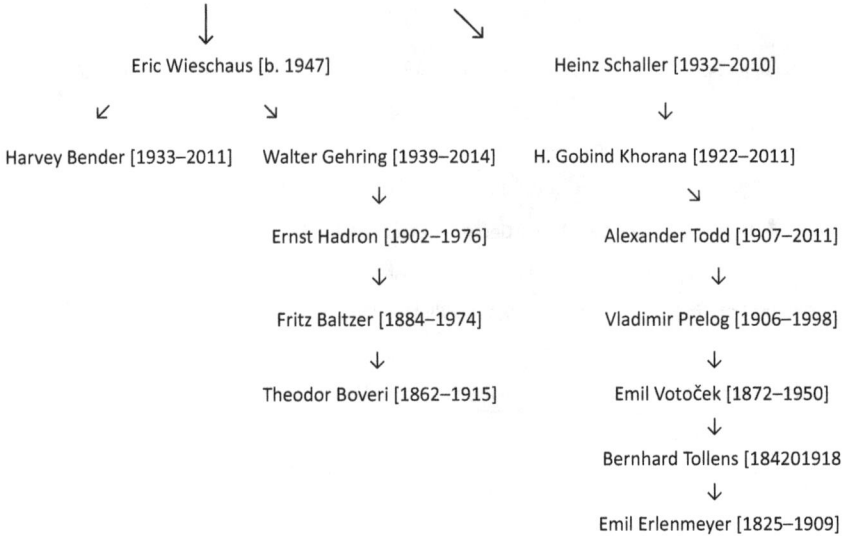

Eric Wieschaus [b. 1947]   Heinz Schaller [1932–2010]

Harvey Bender [1933–2011]   Walter Gehring [1939–2014]   H. Gobind Khorana [1922–2011]

Ernst Hadron [1902–1976]   Alexander Todd [1907–2011]

Fritz Baltzer [1884–1974]   Vladimir Prelog [1906–1998]

Theodor Boveri [1862–1915]   Emil Votoček [1872–1950]

Bernhard Tollens [184201918

Emil Erlenmeyer [1825–1909]

Legend for Pedigree of Christiane Nüsslein-Volhard

Christiane Nüsslein-Volhard (b. 1942) was born in Magdeburg, Germany, and received her PhD in 1974 on RNA polymerase and DNA interaction. She earned a Nobel Prize for her work that demonstrated genes for early fruit fly larval embryos used in body segmentation.

Eric Wieschaus (b. 1947) was born in South Bend, Indiana, and received his PhD in embryology at Yale. His father was a chemical engineer. He isolated lethal mutations that affected developmental stages in larval fruit flies for which he earned a Nobel Prize. His undergraduate mentor was **Harvey Bender** at Notre Dame University.

Harvey Bender (1933–2011) was born in Cleveland, Ohio, and died in South Bend, Indiana. He was a geneticist who studied pseudoallelism in fruit flies.

Walter Gehring (1939–2014) was born in Zürich and died in Basel, both in Switzerland. He first identified in 1983 homeoboxes in animals and plants as conserved developmental genes that initiated segmentation of early embryos.

Ernst Hadorn (1902–2004) was born in Forst and died in Wohler, both in Switzerland. He studied recessive lethal mutations and classified which died before emersion of the first instar larvae and which died at later larval stages or during pupation.

Fritz Baltzer (1884–1974) was born in Zurich and died in Bern, both in Switzerland. He studied developmental genes by using enucleated eggs with a sperm nucleus introduced to identify which events and stages were altered or prevented by such experiments.

Theodor Boveri (1862–1915) was born in Samberg, Germany and died in Würzburg, both in Germany. Boveri used agitation to produce zygotes with different numbers of chromosomes and showed such embryos aborted but different excesses or deficits of chromosomes led to different abnormalities in development and time of death. He concluded that the chromosomes were the carriers of genetic units (genes).

Heinz Schaller (1932–2010). He was a molecular biologist at the University of Heidelberg who worked on virus genes, especially the hepatitis virus.

H. Gobind Khorana (1927–2011) was born in Raipur, Punjab (now Pakistan), and died in Concord, Massachusetts. His parents were Hindus and his father, a tax collector. He received a Nobel Prize for his work on the genetic code and its relation to protein synthesis.

Alexander Todd (1907–1997) was born in Glasgow, Scotland, and died in Cambridge, England. He was a chemist who worked out the composition of nucleotides and cofactor molecules like ATP (adenosine triphosphate) for which he received a Nobel Prize.

Roger J. S. Beer was an organic chemist who taught at the University of Liverpool.

Vladimir Prelog (1906–1998) was born in Sarajevo, Bosnia, and died in Zurich, Switzerland. He was a chemist who worked on alkaloids and other multi-ring compounds, for which he received a Nobel Prize in chemistry.

Emil Votoček (1872–1950) was born in Hostinné in the Austro-Hungarian Empire. He died in Prague, Czechoslovakia. He studied carbohydrates with Tollens and at age 60 he switched to music and wrote about 60 compositions for chamber groups and sonatas.

Bernhard Tollens (1872–1950) was born in Hamburg and died in Gottingen, Germany. He worked with Möbius, Wöhler and Erlenmeyer. His research focused on sugars and other carbohydrates.

Emil Erlenmeyer (1825–1909) was born in Heidelberg and died in Aschaffenburg, Germany. He was an organic chemist and synthesized amino acids. He also introduced the Erlenmeyer flask.

Theophilus Shickel Painter [1889–1969]

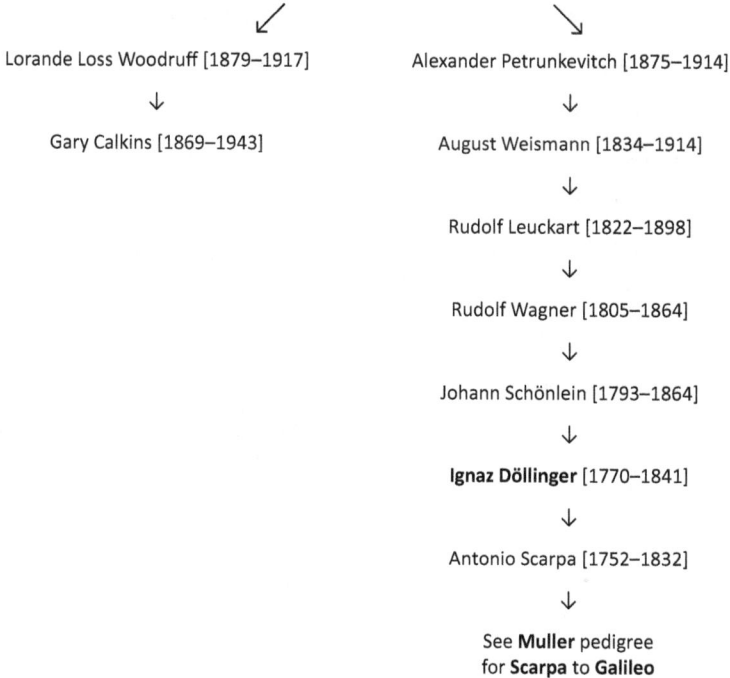

Lorande Loss Woodruff [1879–1917]   Alexander Petrunkevitch [1875–1914]

↓   ↓

Gary Calkins [1869–1943]   August Weismann [1834–1914]

↓

Rudolf Leuckart [1822–1898]

↓

Rudolf Wagner [1805–1864]

↓

Johann Schönlein [1793–1864]

↓

Ignaz Döllinger [1770–1841]

↓

Antonio Scarpa [1752–1832]

↓

See **Muller** pedigree
for **Scarpa** to **Galileo**

Legend for T. S. Painter Pedigree

Theophilus Shickel Painter (1889–1969) was born in Salem, Virginia. His father was a professor of English Literature at Roanoke College, where Painter did his undergraduate BA in 1908. He went to Yale and took an interest in cytology, receiving a PhD by working on spider chromosomes (1913). He went to Germany for a year studying at Würzburg with Theodor Boveri and then enjoying the rest of his time in Naples at its marine station. He joined the faculty at the University of Texas and studied mammalian chromosomes, showing an XY sex determining mechanism in the mammals he studied. He worked with **Muller** on chromosome rearrangements in fruit flies, both spontaneous and X-ray induced. He reported the human chromosome number as 46, 47, or 48 based on testes from sterilized patients declared insane or retarded or executed prisoners. His choice of 48 turned out to be incorrect and 46 replaced Painter's 48 in textbooks. In 1936 Painter worked out a banding pattern in salivary chromosomes of the fruit fly larvae. His method quickly became a basis for studying a variety of chromosomal rearrangements resulting in position effects or complex shifting of chromosome parts. Painter became President of the University of Texas and tried to abide by the university's discrimination barring African Americans from attending. A 1950 Supreme Court nullified the Texas law and the University became integrated.

Alexander Petrunkevitch (1875–1914) was born in Plysky, Russia, and died in New Haven, Connecticut. His father was a politician and young Petrunkevitch excelled in school. He went to Freiburg to get his PhD with **August Weismann** on the cytology of honeybees, demonstrating females were diploid and males were haploid, arising by parthenogenesis from unfertilized eggs. He moved to Yale in 1910 where he worked for the rest of his life studying spiders, both live and fossil forms.

August Weismann (1834–1914) was born in Frankfurt and died in Freiburg, both in Germany. He received an MD at Göttingen but shifted to basic research. He demonstrated that mutilations of tails of rats did not lead to inherited variations. He demonstrated that in flies the reproductive cells (germplasm) were housed separately in the embryo from the soma. He inferred that environmental effects in the soma were not transmitted to the germinal material. He called this the theory of the germplasm.

Rudolf Leuckart (1822–1898) was born in Heinstadt and died in Heidelberg, both in Germany. He was a parasitologist.

Rudolf Wagner (1805–1864) was born in Beyreuth and died in Göttingen, both in Germany. He taught anatomy and discovered the germinal vesicle in which the egg cell is embedded in mammals.

Gary Calkins (1869–1943) was born in Valparaiso, Indiana and died in New York City. He was a protozologist who spent his career at Columbia University studying marine protozoa.

Llorande Woodruff (1879–1917) was born in New York City and died in New Haven, Connecticut. He was a protozologist with an emphasis on ciliates. His work on Paramecium led to the discovery of the genetic basis for autogamy by Sonneborn. He taught at Yale and shifted to history of biology and was one of the first to teach a course in the history of biology.

Johann Schönlein (1793–1841) was born and died in Bamberg, Germany. He showed the cause of ringworm was a fungus, *Trichophyton schönlandii*. He also named tuberculosis for its microscopic pathology replacing the contemporary term, consumption.

Ignaz Döllinger (1770–1841) was born in Bamberg, Germany where his father was a professor and physician. He died in Munich. He got his doctorate in 1794 and studied in Wurzburg, Padua, and Vienna before settling in to Munich where he was Professor of anatomy. His medical degree was from Padua and his doctoral advisor was Antonio Scarpa. Döllinger taught medicine as a natural science. He devoted most of his research to embryonic development. His students included Louis Agassiz, Karl Ernst von Baer, Christian Pander, Lorenz Oken, and Johann Lukas Schönlein. Pander and von Baer helped establish the field of embryology.

Pander introduced the idea of three germ layers in the early embryo — ectoderm, mesoderm, and endoderm. It was von Baer who described development as epigenetic with organs formed not by enlargement but by differentiation of embryonic germ layers. Schönlein was one of the first German professor to lecture in German rather than in Latin. Oken was a founder of a holistic, transcendental approach to biology called *naturphilosophie* that was extended by Goethe. It tried to make sense of homologous structures and how they came to be modified.

Antonio Scarpa (1752–1832) was born in Liorenzaga, Italy and died in Padua, Italy. He received his medical degree at the University of Padua, studying with Giovanni Battista Morgani. He studied the anatomy of the inner ear and cardiac nerves. He taught at the University of Modena. He was a bachelor and had several children out of wedlock. Born poor, he became wealthy, and a collector of art. Scarpa's most noted student was Ignaz Döllinger.

Linus Carl Pauling [1901–1994]

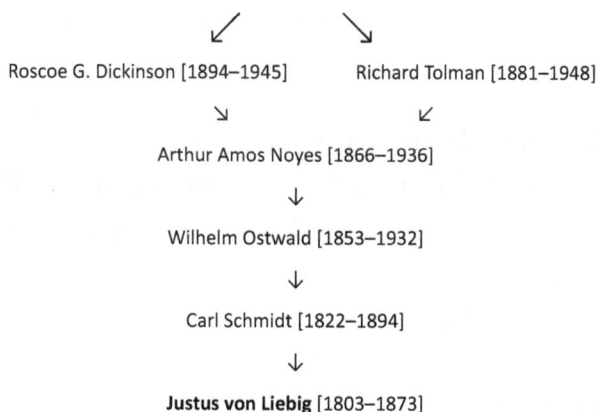

↙   ↘

Roscoe G. Dickinson [1894–1945]   Richard Tolman [1881–1948]

↘   ↙

Arthur Amos Noyes [1866–1936]

↓

Wilhelm Ostwald [1853–1932]

↓

Carl Schmidt [1822–1894]

↓

Justus von Liebig [1803–1873]

Intellectual Pedigree of Linus Pauling

Linus Pauling (1901–1994) was born in Portland, Oregon, and died in Big Sur, California. His father was a pharmacist who died when Linus was 10 years old. Pauling took an interest in chemistry from a childhood friend and majored in chemistry at Oregon State for his BA. He went to Caltech for his PhD (1925) on X-ray diffraction of crystals. His mentors were **Roscoe Dickinson** and **Richard Tolman**, both of whom got their PhDs with **Arthur Amos Noyes**. Pauling later returned to Caltech as a faculty member. He contributed to knowledge of the chemical bond, worked out the alpha-helix in proteins, and showed that sickle cell anemia was due to a single amino acid substitution. He designated sickle cell anemia as a molecular disease. Pauling won two unaccompanied Nobel Prizes, Chemistry and Peace. He had led an international petition of scientists around the world urging a stop of nuclear weapons testing.

Roscoe Gilkey Dickinson (1894–1945) was born in Brewer, Maine, and died in Pasadena, California. He attended MIT for his

BS degree and got his PhD at Caltech in 1920 (the first PhD to be awarded there). He studied the structure of crystals using X-ray diffraction.

Richard Tolman (1881–1948) was born in West Newton, Massachusetts, and died in Pasadena, California. He specialized in statistical mechanics and applied relativity to chemistry. He got his PhD at Caltech. He interpreted electric current as the movement of electrons along a wire.

Arthur Amos Noyes (1866–1936) was born in Newburyport, Massachusetts, and died in Pasadena, California. He got his PhD in Leipzig in 1890. Noyes never married and contributed part of his salary to Caltech to support students in chemistry. He was a gifted teacher and believed the earlier that students experienced research, the more enthusiastic they would be. His research was on chemical solutions using rare earths to analyze their properties.

Wilhelm Ostwald (1853–1932) was born in Riga, Latvia, and died in Großbothen (now Grimma) Germany. He became a chemist and studied reaction velocities, catalysis, and chemical equilibria for which he received a Nobel Prize in 1909. He devised methods for nitrogen fixation, and he introduced the idea of the mole as an element's atomic weight in grams.

Carl Schmidt (1822–1894) was born in Mitau, Latvia, and died in Dorpat, Estonia. He got his PhD in Giessen with **Justus von Liebig**. He studied the chemical composition of the mantel or outer coat of ascidians and showed they were composed of a form of cellulose that he called tunicine.

Justus von Liebig (1803–1873) was born in Darmstadt and died in Munich, both in Germany. He got his PhD in Erlangen and became a founder of organic chemistry. His study of plant nutrients at the University of Giessen led to the bouillon cube and the commercial sale of dietary supplements. It also led to the use of ammonia as a soil fertilizer, greatly increasing crop yields for farmers.

Lionel Penrose [1898–1972]

↓

Sigmund Freud [1856–1939]

↙   ↘

Franz Brentano [1838–1917]   Ernst von Brücke [1819–1892]

↓   ↓

Franz Clemens [1815–1862]   **Johannes Müller** [1801–1858]

↓

See **Gegenbaur** pedigree from
Müller to **Friedrich Leibniz**

Legend for Intellectual Pedigree of Lionel Penrose

Lionel Sharples Penrose (1898–1972) was born and died in London, England. His father and mother were artists. He was raised as a Quaker and in WWI served as a conscientious objector ambulance driver in France. He received his MD in 1928 and in 1930 specialized in psychiatric patients after visiting Freud and his associates in Vienna. In 1930 he studied 1280 patients in institutions for the feebleminded, as they were called then. He classified them into categories: anoxic birth, Down syndrome, excess males with facial dysmorphia, Mendelian likelihood due to cousin marriages, microcephalic, etc. This showed that mental retardation had many causes and a range of mental and physical disabilities. He identified a paternal age effect for autosomal dominant disorders such as achondroplasia and a maternal age effect for Down syndrome. He was a founder of medical genetics in the United Kingdom.

Sigmund Freud (1856–1939) was born in Pribor, Czechoslovakia, and died in London, England. He studied mental illness in France and developed a talk therapy based on a theory of personality dynamics. He called the operating self the ego, the unconscious self the id and associated it with a pleasure-seeking, sexual, and often destructive behavior. Serving as a brake to the id, he invoked the superego as a conscience or regulator of social behavior. He invoked an Oedipal complex stemming from ancestral incest that governs human sexual behavior, much of it repressed or diverted into symbolic forms. His views were widely accepted until the 1970s when they were replaced by pharmacological approaches to regulating mental illness or neurosis.

Franz Brentano (1838–1917) was born in Marienberg am Rhein, Germany, and died in Zürich, Switzerland. He became a Catholic priest and obtained a PhD in philosophy studying Aristotle's theory of how the senses work. He opposed papal infallibility and left the priesthood for an academic life as a philosopher and psychologist. He taught at the University of Vienna.

Ernst Wilhelm von Brücke (1819–1892) was born in Berlin, Germany, and died in Vienna, Austria. He taught Freud neuroanatomy and impressed positivism on his students, arguing that all phenomena can be explained by science and there is no need to invoke vitalism for living phenomena. He also felt art should not be limited to mimicking reality but had aesthetic standards often involving non-realist impressions expressed as art.

Franz Clemens (1815–1862) was born in Coblenz, Germany, and died in Rome, Italy. He was a Jesuit philosopher who taught at Bonn and later at Münster. His doctorate was on Giordano Bruno and Nicholas of Cusa. He was an apologist for the Church on all its theological positions.

Richard J. Roberts [b. 1943]

↙	↓	↘
Daniel Nathans [1928–1999]	**James D. Watson** [b. 1924]	David Olis
↓	↓	
Fritz A. Lipmann [1899–1986]	See **Muller** pedigree	
↓		
Otto Meyerhof [1884–1951]		
↓		
Otto Warburg [1883–1970]		
↓		
Emil Fischer [1852–1919]		
↓		
Adolph von Baer [1835–1917]		
↓		
Friedrich Kekulé [1829–1896]		

Legend for Intellectual Pedigree of Richard Roberts

Richard J. Roberts (b. 1943) was born in Derby, UK. His father was an auto mechanic. He got his BS and PhD at the University of Sheffield. His dissertation with **David Olis** was on flavonoids. He then spent his postdoctoral years at the Medical Research Center in Cambridge and with **J. D. Watson** at Cold Spring Harbor. Influential in his shift to molecular biology was reading **Kendrew's** papers. He discovered (independently of **Philip Sharp**) the intron and exon composition of genes and realized it required gene splicing mechanisms as the informational exons would contribute to the protein synthesis of the final product. He was a founder of New England Biolabs that specializes in isolating enzymes and preparing sequences for investigators around the world.

Daniel Nathans (1928–1999) was born in Wilmington, Delaware and died in Baltimore, Maryland. His parents were immigrants from Russia. He got his BS in chemistry from the University of Delaware and his MD from Washington University in St. Louis. He then went to the Rockefeller Institute to work in **Fritz Lipmann's** laboratory. He did his research and teaching at Johns Hopkins University. He first found and defined restriction enzymes.

Fritz Albert Lipmann (1899–1986) was born in Königsberg, Germany and died in Poughkeepsie, New York. He got his MD at Humboldt University in Berlin and his PhD with **Otto Meyerhof** at Ludwig Maximillian University in Munich. In 1931 he became a fellow at the Rockefeller Institute working with **Philip Levine**. He worked on carbohydrate metabolism and the role of ATP in that glycolytic process. He remained in the United States as a professor at Harvard.

Otto Meyerhof (1884–1951) was born in Hanover, Germany and died in Philadelphia, Pennsylvania. He was raised in a wealthy family and chose medicine. He attended Heidelberg University and Strasbourg for his medical education. He was mentored by **Otto Warburg** and stayed in Germany until 1938 when Nazi racial laws forced him to leave Germany. He went to Paris and studied glycolysis and isolated ATP. He showed glycogen (an animal carbohydrate made in the liver) was broken down to lactic acid in muscle cells. When the Nazis invaded France, he was helped in his escape by the Unitarian Service Committee. He worked at the University of Pennsylvania for the rest of his career.

Philip Sharp [b. 1944]

↙   ↓   ↘

Frederick Sanger [1918–2013]   Victor Bloomfield [b. 1938]   James Watson [b. 1928]

↓   ↓   ↘

Albert Neuberger [1908–1996]   Robert Alberty [1921–2014]   Norman Davidson

↓   ↓   ↓

Charles Harrington [1897–1972]   Farrington Daniels [1889–1972]   Linus Pauling [1901–1994]

↓   ↓   ↓

George Barger [1878–1939]   Theodore W. Richards [1831–1928]   See pedigree

↓

John Parsons Cooke [1827–1894]

↓

Henri Victor Regnault [1810–1878]

↓

Justus von Liebig [1803–1873]

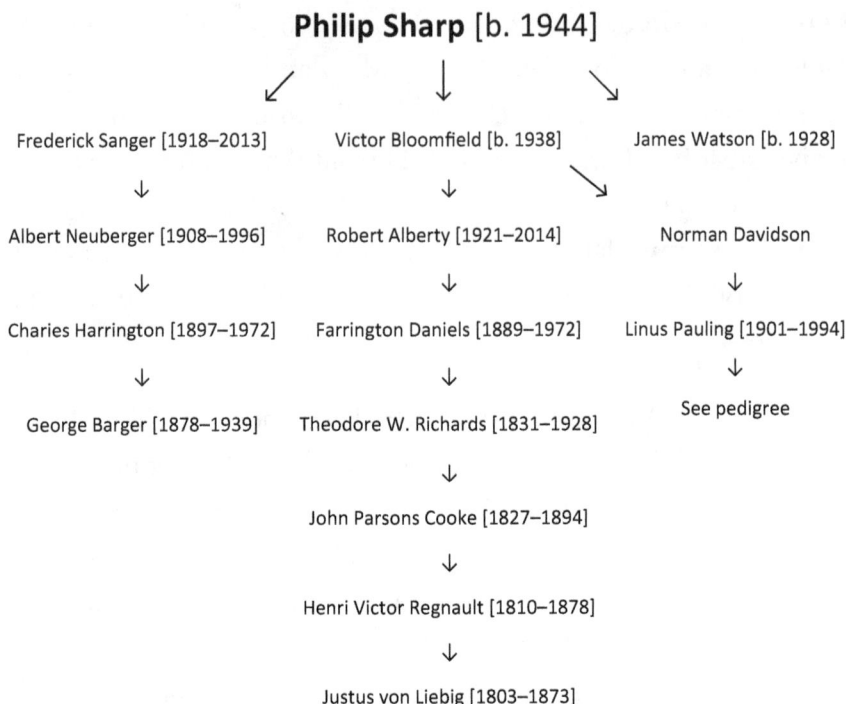

Phillip Sharp [b. 1944] was born in Falmouth, Kentucky, and grew up in a farm. He attended Union College and the did graduate work in chemistry at the University of Illinois in Urbana under the mentorship of Victor Bloomfield. He did his postdoctoral work at Caltech studying plasmids (movable elements). Independently of Richard Roberts, he found split genes (introns and exons). He identified about 60 mentors or collaborators in his career.

Frederick Sanger (1918–2013) was born in Rendcomb and died in Cambridge both in England. His father was an MD. Both parents died while he was an undergraduate at Cambridge. He was raised a Quaker and was a conscientious objector in WWII. He worked on the structure of insulin and devised a technique for sequencing DNA. These two discoveries led to two Nobel Prizes in chemistry.

Albert Neuberger (1908–1996) was born in Hassfurt, Germany and died in Hampstead, England. His father was a cloth merchant and jewish and when Hitler came to power, his family went to London. He got his MD in Wurzburg and liked biochemistry. In London, he worked on glycoproteins and showed that most proteins had a carbohydrate attachment. He also worked on porphyrins and lectins.

Charles Harington (1897–1970) was born in Llanerfyl, Wales, and died in London, England. His father was a Minister. He received his PhD in 1922 studying protein metabolism. He isolated and synthesized thyroxine and glutathyroid.

George Barger (1878–1839) was born in Manchester, England and died in Aeschi, Switzerland. His father was Dutch and his mother was English. His father was an engineer. Barger became a chemist and studied alkaloids, extracting them from ergot. He also studied vitamin B.

Victor Bloomfield (b. 1938) was born in Newark, New Jersey. He got his PhD with Robert Albert. He has spent most of his career at the University of Minnesota. He has studied the physical chemistry of nucleic acids.

Norman Davidson (1916–2002) was born in Chicago and died in Pasadena. He was a Rhodes scholar and got his PhD at the University of Chicago. At Caltech, he studied the renaturing of dissociated DNA and helped worked out techniques for genetic engineering and sequencing of DNA by using its physical chemical properties. He was a recipient of the National Medal of Science.

Linus Pauling (1901–1994) was born in Portland Oregon, and died in Big Sur, California. His father was a pharmacist who died when Pauling was 10 years old. Pauling took an interest in chemistry from a childhood friend and majored in chemistry at Oregon state for his BA. He went to Caltech for his PhD (1922) on X-ray diffraction of crystals. His mentors were Roscoe Dickenson and Richard Tolman, both of whom got their PhD from Arthur Amos Noyes. Pauling later returned to Caltech as a faculty member. He contributed to knowledge of the chemical bond, worked out the alpha-helix in proteins, and showed sickle cell anemia was due to a single amino acid substitution. He designated sickle cell anemia as a molecular disease. Pauling won two unaccompanied Nobel Prizes in Chemistry and in Peace. He had led an internaional petition of scientists around the world urging a stop of nuclear weapons testing.

James Dewey Watson (b. 1928) was born and raised in Chicago, Illinois. His memoir, *The Double Helix* (1968) became a best seller. His intial interest was in ornithology, but Fernandus Payne told Watson that if he wanted to go into that field, he should apply to Cornell University. If he wanted first rate genetics, he should come to Indiana University. Payne was a student of Morgan and Wilson and active in the IU Zoology Department when Watson came to Indiana University for his PhD. At the time, Indiana University had four world class geneticists — Muller using fruit files, Sonneborn using Paramecia, Luria using viruses, and Clenand using Oenothera. Watson chose Luria for his dissertation work because he felt the study of virus genetics was most likely to reveal gene action and composition. He also felt Muller's courses stressing the gene were the most influential in shaping his career. His work with Francis Crick at Cambridge led to their 1953 papers in *Nature* that provided the chemical structure of DNA and the biological implications of that structure for gene replication, mutation, and function. Through Luria, Watson became a member of the phage group that Delbrück founded when he went to Caltech after leaving Germany and settling into the United States. Delbrück's quest for the physical structure of the gene resonated with Watson. Watson also helped launch the Genome Project. Watson's personality was controversial because he spoke his mind on issues with a tart tongue causing responses in kind. He believed that human intelligence was tied to heredity and this made him a racist in the interpretation of some of his critics. He said his son was schizophrenic and he loved his son but recognized genetic components contributed to that condition. He was accused of being a sexist in his treatment of Rosalind Franklin in his memoir, but he said he was reflecting how males of his generation reacted to females but never doubted her competence as a scientist. He denied that seeing her photo of DNA enabled him to figure out the structure of DNA. He said it confirmed what he and Crick were already considering — a double helix for DNA.

Robert Alberty (1921–2014) was born in Winfield, Kansas and died in Cambridge, Massachusetts. He was a biophysical chemist, receiving his PhD at the University of Wisconsin. He was Dean at MIT.

Farrington Daniels (1889–1972) was born in Minneapolis. He worked on solar energy and the Manhattan Project, but after WWII, he became an opponent of nuclear weapons development and was a founder of the *Bulletin of Atomic Scientists*.

Theodore Richards (1831–1928) was born in Germantown, Pennsylvania and died in Cambridge. Massachusetts. His father was an artist. He worked out most of the atomic weights of the elements and was the first American to win a Nobel Prize in Chemistry.

John Parsons Cooke (1827–1894) was born in Boston, Massachusetts. He worked on atomic weights.

Henry Victor Regnault (1810–1878) was born in Aachen, Germany and died in Paris, France. He worked on atomic weights.

Jean Baptiste Dumas (1800–1884) was born in Alès and died in Cannes, both in France. He studied botany (with A. P. de Candolle) and chemistry (with Pierre Provost). He identified the kidneys as the source for urea excretion. He used vapor densities to determine atomic weights.

Pierre Provost (1751–1839) was born and died in Geneva. He father was a pastor. He was a philosopher, economist, and physicist. He demonstrated that all material bodies emit heat.

Justus von Liebig (1803–18730) was born in Darmstadt and died in Munich, both in Germany. He was a founder of organic chemistry and made many contributions to agricultural chemistry, including the use of nitrogen fertilizers for soil and the development of the bouillon cube as a food that can be stored and used for soups and other cooking. He is also the grandfather of Max Delbrück.

Carl Sagan [1934–1996]

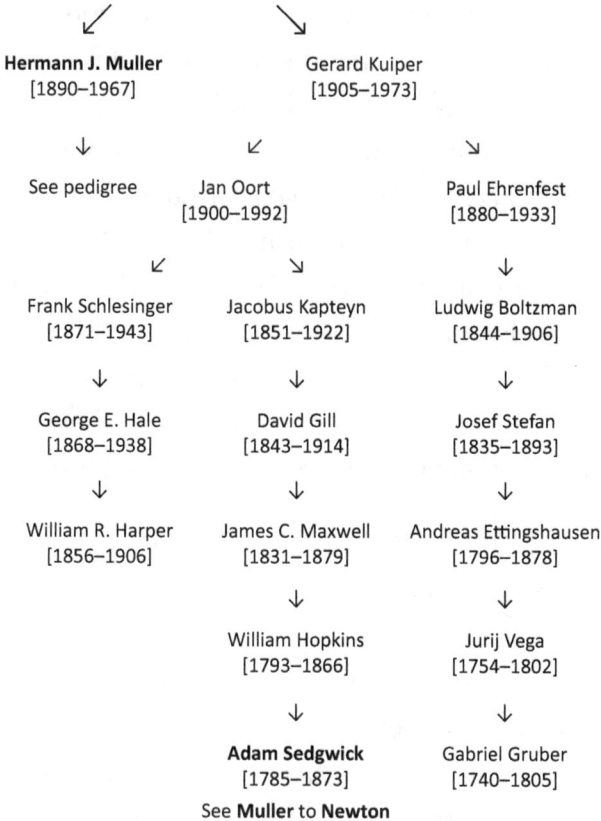

↙   ↘

Hermann J. Muller
[1890–1967]

Gerard Kuiper
[1905–1973]

↓   ↙   ↘

See pedigree

Jan Oort
[1900–1992]

Paul Ehrenfest
[1880–1933]

↙   ↘   ↓

Frank Schlesinger
[1871–1943]

Jacobus Kapteyn
[1851–1922]

Ludwig Boltzman
[1844–1906]

↓   ↓   ↓

George E. Hale
[1868–1938]

David Gill
[1843–1914]

Josef Stefan
[1835–1893]

↓   ↓   ↓

William R. Harper
[1856–1906]

James C. Maxwell
[1831–1879]

Andreas Ettingshausen
[1796–1878]

↓   ↓

William Hopkins
[1793–1866]

Jurij Vega
[1754–1802]

↓   ↓

Adam Sedgwick
[1785–1873]

Gabriel Gruber
[1740–1805]

See **Muller** to **Newton**

Carl Sagan (1934–1996) was born in Brooklyn, New York and died in Seattle, Washington. His parents were immigrant reform Jews who moved to New Jersey where young Sagan attended high school. He was precocious and accepted to the University of Chicago where he studied astronomy. He enjoyed Urey's work with abiotic synthesis of molecules found in living systems and Kuiper's studies of the solar planets. He got his PhD with **Kuiper** and took a job at Harvard but did not make tenure, in part because he spent more time popularizing science than doing peer reviewed contributions to astronomy. Sagan was hired by Cornell University and spent the rest of his career there. He was married three times and his first wife, **Lynn Margulis**, was an evolutionary biologist who provided evidence for an endosymbiotic origin of cell organelles like mitochondria and chloroplasts. Sagan's books include *The Cosmic Connection, The Dragons of Eden*, and *The Pale Blue Dot.* Sagan showed that the hot atmosphere of Venus was a consequence of a greenhouse effect of its gases. He also argued that climate change due to greenhouse gases in the earth's atmosphere would lead to massive changes in life on earth if fossil fuels were not curtailed. His *Cosmos* series of television shows on science motivated many young viewers to enter science for their careers.

Gerard Kuiper (1905–1973) was born in Tuitjenhorn, Netherlands and died of a heart attack while in Mexico City, Mexico. His father was a tailor. He had unusually good eyesight and thought this would be useful for an astronomer. His PhD in astronomy at Leiden was on binary stars. He shifted to planetary astronomy and discovered a moon on Uranus and a moon on Neptune. He showed Mars had CO_2 in its atmosphere and methane was present on Titan (a moon of Saturn). He predicted the Kuiper belt which extends past Neptune and it is 20 times wider than the asteroid belt between Mars and Jupiter.

Jan Oort (1900–1992) was born in Franeker, Netherlands and died in Leiden, Netherlands. His father was a physician. He predicted the Oort cloud as a region near the distant focus of comets orbiting the sun. He also proved that the Milky Way rotates and in 1932 that most of the universe is composed of dark matter.

Paul Ehrenfest (1880–1933) was born in Vienna and died in Amsterdam. His father was a grocer and an observant Jew, but

Paul became an atheist. The courses he took with **Boltzmann** converted him to physics. He helped develop quantum mechanics and became a friend of Einstein. He was an excellent teacher and mentored **Samuel Goldschmidt** and **Enrico Fermi**. He suffered from depression and shot his youngest son (who had Down syndrome) and himself.

Frank Schlesinger (1871–1943) was born and died in New York City. He did his undergraduate work at CCNY and his PhD at Columbia University in 1898. He developed photographic methods for astronomy and was a director of the Ukiah California observatory. He also taught at Pittsburgh studying parallax of stars, binary stars, and produced the criteria for interpreting photographic plates used in mapping the Milky Way.

Jacobus Kapteyn (1851–1922) was born in Amsterdam and died in Barneveld, both in the Netherlands. He noted that stars seemed to move in two streams in opposite directions providing evidence for the rotation of the Milky Way. He devised criteria for mapping stars in the galaxy, including direction, luminosity, and mass.

Ludwig Boltzmann (1844–1906) was born in Vienna and died in Duino, Italy. His father was a revenue official and died when his son was 15. He enjoyed physics and was mentored by **Josef Stefan** for his PhD. He contributed to the analysis of atomic properties, including their mass, charge, and location led to the physical attributes of matter as elements or molecules, including viscosity, diffusion, and conductivity. His students included **Svante Arrhenius** and **Walther Nernst**. He suffered from depression and hanged himself in 1906.

George Ellery Hale (1868–1938) was born in Chicago and died in Pasadena. His father was wealthy as an installer of elevators after the Chicago fire of 1871. Hale was a prodigy and built his own telescope when he was 14. He went to MIT and to Berlin before serving as director of the Yerkes, then Mt. Wilson, and then Palomar observatories. He discovered solar vortices in the sun's atmosphere, the magnetic properties of sunspots, and the cause of the disappearance and reappearance of sunspots at predicted locations on the sun. Most of his adult life he suffered from severe headaches, depression, and insomnia.

David Gill (1843–1914) was born in Aberdeen, Scotland and died in London. His father was a watchmaker. Gill was taught by **James Maxwell** at Aberdeen University. He studied solar parallax, a transit of Venus across the sun, and contributed to an inventory of the skies, providing the locations of an estimated half million stars.

Josef Stefan (1835–1893) was born in Klagenfurt and died in Vienna, both being in Austria. His father was a milling assistant and his mother a housemaid. He was mentored in mathematics and physics by **Andreas von Ettingshausen** and received his PhD. He studied black body radiation and applied his mathematical calculations to estimate the sun's surface temperature and radiated heat. The temperature was 5430 degrees Centigrade.

James Clerk Maxwell (1831–1879) was born in Edinburgh, Scotland and died in Cambridge, England. His father was a lawyer. His mother died when he was 8 years old. He was a gifted student and got his PhD at Cambridge with **William Hopkins**. Maxwell showed that electricity and magnetism were related and demonstrated how a moving magnet could induce an electric current in a wire. He also determined that the current flowed at a rate approximating that of the speed of light. He showed that Saturn's rings had to be composed of small rocks rather than a solid sheet. His equations relating electricity and magnetism elevated him to the status of being the third most noted of modern physicists.

William Hopkins (1793–1866) was born in Kingston-on-Soar, England and died in Cambridge, England. He demonstrated that the melting point of a compound or element increases with increasing pressure. He had many outstanding students including **Kelvin, Maxwell**, and **Galton**. He received his MA under the mentorship of **Adam Sedgwick**. Sedgwick introduced him to geology and Hopkins used applied math to describe the proprieties of crystals. In his last years, Hopkins was confined to a lunatic asylum suffering from mania.

Adam Sedgwick (1785–1873) was born in Dent, England and died in Cambridge. His most famous student was **Charles Darwin**. He had invited Darwin to go on a field trip to study rock strata in Scotland. Sedgwick described the Devonian and Carboniferous strata in Great Britain. While he was a friend of Darwin's in later life, he ridiculed the idea of evolution by natural selection, claiming it was "a dish of rank materialism cleverly cooked and served up."

Jurij Vega(1754–1802) was born in Zagorica pri Dolskem in Slovenia and died near Vienna, Austria. He was educated at Ljubljana and served in the artillery during the Napoleonic wars. He wrote books on logarithms and trigonometric functions. He taught mostly at Mainz. He was mentored by **Gabriel Gruber**.

Gabriel Gruber (1740–1805) was born in Vienna, Austria and died in St. Petersburg, Russia. He was a Jesuit with multiple interests and talents including mathematics, engineering, shipbuilding, navigation, painting, and architecture. He helped shape education in Russia. He died at home in a fire.

Erwin Schrödinger [1867–1961]

↓

Friedrich Hasenöhrl [1874–1915]

↙   ↘

Franz Exner [1849–1926]	Ludwig Boltzmann [1844–1906]
↓	↓
Wilhelm Röntgen [1845–1923]	Josef Stefan [1835–1893]
↓	↓
August Kundt [1839–1894]	Andreas von Ettingshausen [1796–1878]
↓	↓
Heinrich Magnus [1802–1870]	Ignaz Lindner [1777–1835]
↓	

Eilhard Mitscherlich [1794–1806]

↓

Friedrich Strohmeyer [1776–1835]

↓

Johann Gmelin [1748–1804]

Legend for Erwin Schrödinger

Erwin Schrödinger (1867–1961) was born and died in Vienna, Austria. He was a physicist and philosopher. He won a Nobel Prize for his work on wave mechanics applied to quantum theory. He also wrote a best seller, (What is Life?), that helped launch the field of molecular genetics. He based his book on a paper written by **Max Delbrück**. Schrödinger was unconventional. He had a wife and mistress and they lived together as a three-some. He also fathered two Irish children out of wedlock. He liked Eastern mysticism.

Wilhelm Röntgen (1845–1923) was born in Lennep and died in Munich, both in Germany. He is best known for discovering X-rays for which he received a Nobel Prize.

Ludwig Boltzmann (1844–1906) was born in Vienna, Austria, and died in Trieste, Italy. He was a physicist known for his work on the electric conductivity of gases. He suffered from depression and committed suicide.

August Kundt (1839–1894) was born in Schwerin and died in Lübeck, both in Germany. He is known for his work on polarization of light and the velocity of particles suspended in gases. He applied his findings to electrochemistry.

Heinrich Magnus (1802–1870) was born and died in Berlin, Germany. He discovered the element tellurium.

Friedrich Strohmeyer (1776–1835) was born and died in Göttingen. He discovered the element cadmium.

Eilhard Mitscherlich (1794–1863) was born in Wilhelmshaven and died in Berlin, both in Germany. He was a chemist and studied arsenical and phosphate compounds.

Johann Gmelin (1784–1804) was born in Tübingen and died in Göttingen, both in Germany. He was a founder of the field of ornithology and completed the 13th edition of Linnaeus's *Systema Naturae* classifying several hundred species of birds.

Josef Stefan (1835–1893) was born in St. Peter and died in Vienna, both in Austria. His father was a mill worker and his mother was a domestic servant. He attended the University of Vienna and learned mathematics and physics from Ettingshausen. He worked out the physics and mathematics of black body radiation the physics of heat on gases.

Andreas von Ettingshausen (1796–1878) was born in Heidelberg, Germany and died in Vienna, Austria. He was a mathematician and physicist. He devised the first electric run machine and contributed to combinatorial analysis and the solution to binomial coefficients.

Ignaz Lindner (1777–1835) was an Austrian mathematician and military officer. He taught mathematics at the University of Vienna.

George Harrison Shull [1874–1954]

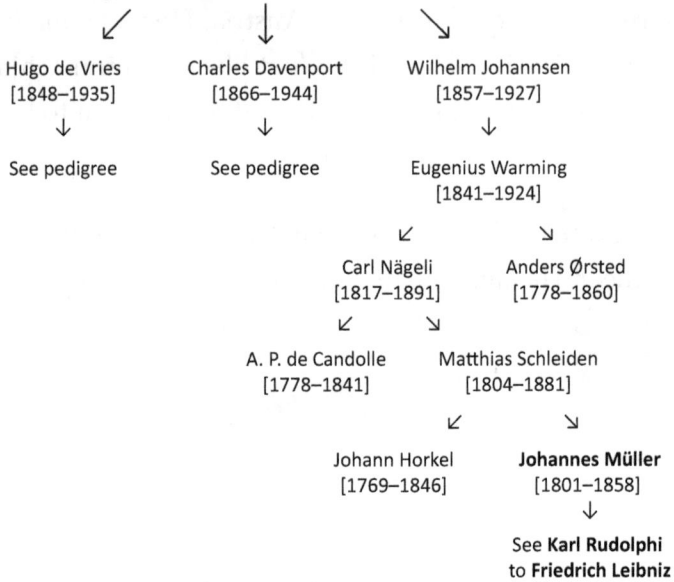

Hugo de Vries	Charles Davenport	Wilhelm Johannsen
[1848–1935]	[1866–1944]	[1857–1927]
↓	↓	↓
See pedigree	See pedigree	Eugenius Warming
		[1841–1924]

Carl Nägeli   Anders Ørsted
[1817–1891]   [1778–1860]

A. P. de Candolle   Matthias Schleiden
[1778–1841]   [1804–1881]

Johann Horkel   **Johannes Müller**
[1769–1846]   [1801–1858]
                        ↓

See **Karl Rudolphi**
to **Friedrich Leibniz**

George Harrison Shull (1874–1954) was born on a farm in Clark County Ohio and died in Princeton, New Jersey. His father was a farmer and his mother a horticulturist. Shull went to Antioch College for his BA in 1901 and to the University of Chicago for his PhD in 1904. He then took a job at the new Biological Station established by **Davenport** at Cold Spring Harbor in NY. He worked on Oenothera mutations and compared them to mutations in maize (corn). He noted that inbred strains of corn (by self-pollination) led to weaker plants in size and number. When he crossed different inbred strains, they produced a hybrid vigor and uniform quality of maize kernels. This work was published in 1908. In 1914 he called hybrid vigor by the term heterosis. It would take another ten years before hybrid corn became a commercial enterprise enormously benefitting farmers planting hybrid corn. In 1915 he became a professor at Princeton. In 1916 he founded the journal, *GENETICS*.

Wilhelm Johannsen (1857–1927) was born and died in Copenhagen, Denmark. His father was an army officer and could not afford to send his son to college. He was apprenticed as a pharmacist and learned as much as could on his own. In his last year he had saved enough to spend one year taking botany courses at the University of Copenhagen and was captivated by **Warming's** lectures. He worked at the Carlsburg Laboratories and studied dormancy in plants, using agents to induce germination. He began his studies of beans about 1900 and showed pure lines had a constant genotype and that selection could establish pure lines. Variation within a pure line was caused by the numerous environments encountered by a growing population of a pure line. His work was influential to classical genetics which found through Johannsen the contribution of numerous genetic factors to establishing quantitative traits. Johannsen introduced the terms genotype, phenotype, and gene to assist a common defined terminology for genetics.

Eugenius Warming (1841–1924) was born in Wando, Denmark and died in Copenhagen, Denmark. He received his PhD in botany at the University of Copenhagen in 1871. He studied plant distribution and adaptations in Greenland, Venezuela, and the West Indies, describing his findings as "phytogeography" and later plant "amfund" or plant communities. He is recognized as an early founder of plant ecology. His most noted student was **W. Johannsen** with whom he co-authored a book on plant physiology.

Carl Nägeli (1817–1891) was born in Zurich, Switzerland, and died in Munich, Germany. He received his MD in Zurich and his PhD in Botany at Berlin in 1840. His mentors were **A. P. de Candolle** and **Matthias Schleiden**. He identified the genetic component of plants as idioplasm and named meristem, phloem, and xylem in his analysis of plant anatomy. He and **Gregor Mendel** independently worked on *Hieracium* crosses and both recognized that it differed from Mendel's work in *Pisum*. *Hieracium* had a complex heredity, many of its species reproduced by partheno-genesis but needed the stimulus of pollen to induce it. A smaller percent in such species also produced occasional Mendelian results. Nägeli and Mendel exchanged correspondence describing their results and sent seeds to each other to farther extend this work. The reproductive process, called apomixis, was not worked out until 1898.

Augustin P. de Candolle (1778–1841) was born and died in Geneva, Switzerland. He was of Huguenot ancestors who fled to Switzerland. He was a botanist and first described a biological clock for plants. He also described "nature's war", a view that Dar-wain incorporated in his theory of natural selection.

Pierre Vacher (1763–1841) was born and died in Geneva, Swit-zerland. He was a pastor and taught botany. He first equated con-jugation in algae with a sexual process.

Anders Ørsted (1778–1860) was born in Rudkøping and died in Copenhagen, Denmark. He studied nematode worms in soils, explored the Flora of Central, America, and worked out secondary hosts for rust fungi.

Matthias Schleiden (1804–1881) was born in Hamburg and died in Frankfurt, Germany. His father was an MD. Schleiden stud-ied law and became depressed. He survived a gunshot wound to the brain. He took up botany on his uncle Johann Horkel's rec-ommendation and taught and popularized botany, writing a book *The Plant, A Biography*. He studied plant structure and concluded all plants were composed of cells. He teamed up with Theodore Schwann who found the same in animal cells and they proposed the cell theory of eukaryotic organisms.

Johannes Müller (1801–1858) was born in Koblenz and died in Berlin, Germany. His father was a shoemaker and young Muller was considering becoming a priest but was excited by natural history and shifted to medicine and biology. He was a skilled anatomist (the Müllerian ducts are named for him) and had many prominent students, including Rudolph Virchow and Hermann von Helmholtz.

Johann Horkel (1769–1864) was born in Fehmarn and died in Berlin, both in Germany. He was Schleiden's uncle and inspired him to study botany. Horkel studied physiology.

Andreas von Ettingshausen (1796–1878) was born in Heidelberg, Germany and died in Vienna, Austria. He taught at the University of Vienna. He was a physicist and mathematician. He was the first to use electricity to power a machine. He worked out combinatorial mathematics and solutions for binomial coefficients. He was Gregor Mendel's teacher.

Tracey Morton Sonneborn [1905–1981]
↓

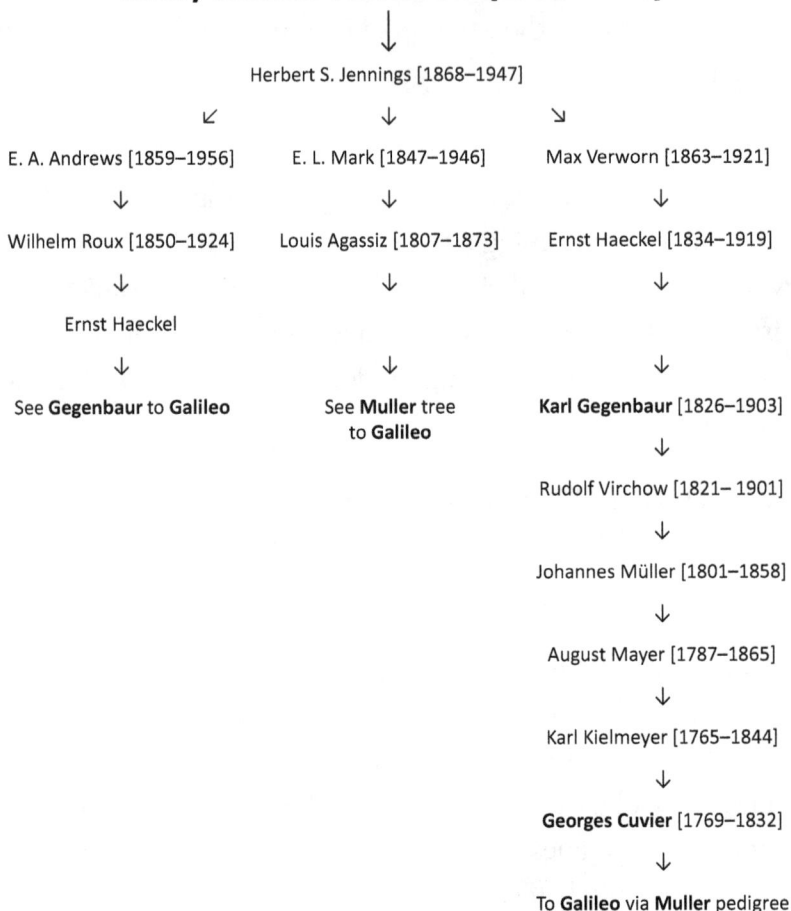

Herbert S. Jennings [1868–1947]

↙　　　　　　↓　　　　　　↘

E. A. Andrews [1859–1956]　　E. L. Mark [1847–1946]　　Max Verworn [1863–1921]

↓　　　　　　↓　　　　　　↓

Wilhelm Roux [1850–1924]　　Louis Agassiz [1807–1873]　　Ernst Haeckel [1834–1919]

↓　　　　　　↓　　　　　　↓

Ernst Haeckel

↓　　　　　　↓　　　　　　↓

See **Gegenbaur** to **Galileo**　　See **Muller** tree to **Galileo**　　Karl Gegenbaur [1826–1903]

↓

Rudolf Virchow [1821– 1901]

↓

Johannes Müller [1801–1858]

↓

August Mayer [1787–1865]

↓

Karl Kielmeyer [1765–1844]

↓

Georges Cuvier [1769–1832]

↓

To **Galileo** via **Muller** pedigree

Commentary on Tracy Sonneborn Pedigree

Tracy Morton Sonneborn (1905–1981) was born in Baltimore, Maryland, and died in Bloomington, Indiana. His family was Jewish and his cousin, **Louis Bamberger**, founded Princeton's Institute for Advanced Study. As an undergraduate he was inspired by lectures of **Ethan Allen Andrews**. For his PhD at Johns Hopkins in 1887, he was mentored by **Herbert S. Jennings**.

He studied flatworms and ciliates. He identified mating types and related them to the reproductive process in Paramecia. A cytoplasmic particle that Sonneborn called a plasmagene turned out to be a bacterial endosymbiont. Sonneborn used cell surgery to rotate a strip of cilia in Paramecia and the cilia continued to beat in their original direction rather than in the new alignment. As a teacher he was inspiring and long advocated to his graduate students the importance of studying nucleocytoplasmic relations.

Herbert S. Jennings (1868–1947) was born in Tonica, Illinois and died in Santa Monica, California. He got his BS at Michigan and his PhD at Harvard in 1896 studying rotifers. He replaced **Brooks** when Brooks retired at Johns Hopkins. After retiring he moved to California and taught at UCLA. He worked out conjugation in Paramecia. His 1930 book, *The Biological Basis of Human Nature*, was a best seller. He was asked to lead an investigation of the eugenics program and wrote a negative appraisal of the Eugenic Record Office which was closed after the report was submitted.

Edward Laurens Mark (1847–1946) was born in Hamlet, NY and died in Cambridge, Massachusetts. He got his BS at the University of Michigan and received his PhD in 1876 in Leipzig with **Leuckart**. He studied the cytology, histology, and anatomy of *Limax* (a slug) and was hired by Harvard where he worked for the rest of his life. He correctly identified the polar body of meiotic cell division as an aborted egg. **Davenport, Castle**, and **Jennings** were his best-known students.

Rudolf Leuckart (1822–1898) was born in Helmstedt, Germany and died in Leipzig, Germany. He was a parasitologist and worked out the life cycles of tapeworms (*Taenia*) and *Trichinella* (a nematode parasite of pigs and humans). His work led to mandatory inspections of meat in Germany and later most industrialized countries.

Ignaz Döllinger (1770–1841) was born in Bamberg, Germany and died in Munich, Germany. His father was a MD and professor. Döllinger believed medicine should be a science as well as an art of healing. He went to Italy to learn the sciences of human anatomy and physiology. He brought back that learning to Germany and among his students were **Agassiz, von Baer, Schönlein, Oken,** and **Pander**.

Louis Agassiz (1807–1873) was born in Môtier, Switzerland and died in Cambridge, Massachusetts. His father was a Huguenot pastor. Agassiz was educated in Erlangen for his PhD (with **Döllinger** and **Carl von Martius**) and got a medical degree in Munich. He was mentored in Paris by **Cuvier** and **Alexander Humboldt**. He described and named the ice age from his studies of glaciers in the Alps and collected fossil fish and classified them before taking a position at Harvard to bring European scholarship to American graduate education. He worked on the distribution and classification of fish in the Western Hemisphere. He believed there were a series of floods that wiped out prevailing species and these were replaced by new acts of creation.

Carl von Martius (1794–1868) was born in Erlangen and died in Munich, Germany. His father was an apothecary. He studied botany and received his PhD at Erlangen. He specialized in palms, going on collecting expeditions in the Amazon and east coast of Brazil.

Lewis Stadler [1896–1954]

↓

Rollins Emerson [1873–1947]

↓

Edward Murray East [1879–1938]

↓

Cyril G. Hopkins [1866–1919]

↙   ↘

George C. Caldwell [1834–1907]   Bernhard C. G. Tollens [1841–1918]

↓   ↓

Friedrich Wöhler [1800–1882]   Emil Erlenmeyer [1825–1909]

↓   ↓

See **Wöhler** to **Linnaeus**   **Justus von Liebig** [1803–1873]
in **McClintock** pedigree   ↓

                            See **Liebig** to **Cassini**
                            in **Brenner** pedigree

Legend for Lewis J. Stadler Pedigree

Lewis John Stadler (1896–1954) was born in St. Louis, Missouri. His father was from Prague, Czechoslovakia and most of his uncles were Rabbis. Stadler's father, however, became a banker and the young Lewis spent two summers on a farm which started his interest in agriculture. Stadler was a "late bloomer" and he was not a good student. His grades were mediocre rather than failing. He took one year of undergraduate work at the University of Missouri and then dropped out. He went into the army during WWI and then got his BA at the University of Florida. He went to Cornell but was considered "indifferent" by **R. A. Emerson** and transferred to the University of Missouri for his PhD in agronomy. It was as a postdoctoral student at Harvard with **E. M. East**, that he finally found in genetics a field that excited his imagination and allowed him to flourish. He studied gene mutations in maize and demonstrated quantitative inheritance. Independent of **H. J. Muller** he successfully induced gene mutations in barley and wheat, but Muller's publication came out earlier. His analysis of the induced mutations in maize with their corresponding natural occurrences as gene mutations demonstrated that X-rays produce more minute rearrangements in or near or deleting normal genes than do spontaneously arising mutations. Stadler died of leukemia while still active in his career at the University of Missouri.

Rollins A. Emerson (1873–1947) was born in Pillar Point, NY and died in Ithaca, NY. His family moved to Nebraska when he was five years old and Emerson got his BA at the University of Nebraska in 1897. He went to Harvard to study with **E. M. East**, getting his PhD in 1913. His work on maize genetics led to his appointment at Cornell University where he taught and did research for the rest of his academic career. He led the maize group at Cornell, identifying ten linkage groups and mapping all known genes of maize.

Cyril George Hopkins (1866–1919) was born in Chatfield, Minnesota and died overseas in Gibraltar. He was raised in South Dakota and studied agricultural chemistry at the University of South Dakota. He got his PhD at Cornell with **George Chapman Caldwell** and then spent a year in Göttingen, Germany with chemist **Bernhard Tollens**.

George Chapman Caldwell (1834–1907) was born in Framingham, Massachusetts and died in Ithaca, New York. He studied botany at Harvard University and went to Heidelberg Germany, spending a year (1885) with chemist **Robert Bunsen**. He then shifted to Göttingen to get a PhD in the laboratory of **Friedrich**

Wöhler (1857). Caldwell taught at Cornell University, Antioch College, and Pennsylvania State University.

Friedrich Wöhler (1800–1882) was born in Hesse-Kassel and died in Göttingen. He received his MD at Heidelberg University and worked with **Leopold Gmelin** who recommended him to go to Stockholm to work with **Jöns Jacob Berzelius**. Wöhler used his chemical knowledge to set up a chemistry laboratory at Göttingen. His most famous work was synthesizing urea from sodium cyanate.

Edward Murray East (1879–1938) was born in Du Quoin, Illinois, and died in Boston, Massachusetts. His father was a mechanical engineer. East got his PhD in 1907 at the University of Illinois. He began in plant chemistry and shifted to plant breeding, discovering multiple factor inheritance in maize. He shifted to the study of heterosis or hybridization and its consequences, applying this to the development of hybrid corn. His book, *Inbreeding and Outbreeding* was a best seller and he became an advocate of the American eugenics movement, favoring restrictive immigration laws and compulsory sterilization of those deemed medically or psychologically unfit to reproduce. He views were racist, sexist, and based on bias.

Bernhard Tollens (1841–1918) was born in Hamburg and died in Göttingen, both in Germany. He worked out the sugar contained in carbohydrates such as starch.

Carl Emil Erlenmeyer (1825–1903) was born in Taunusstein and died in Aschaffenburg, both in Germany. He studied the chemistry of alcohols, aldehydes and introduced the use of double and triple carbon bonds to molecules bearing carbon.

Mankambo Sambasivan Swaminathan (b. 1925)

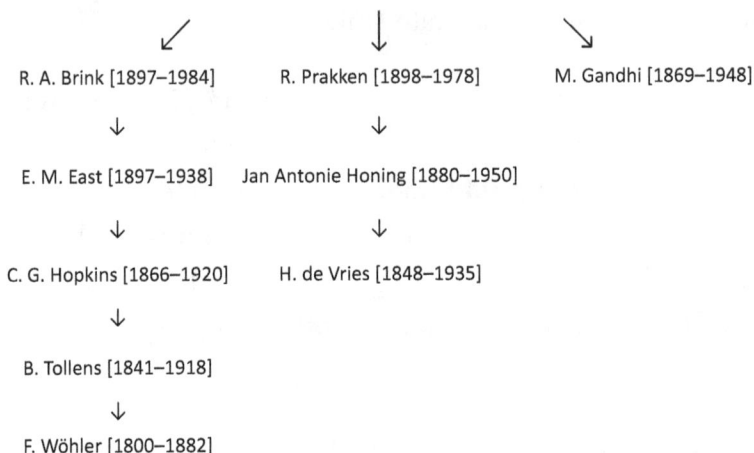

↙	↓	↘
R. A. Brink [1897–1984]	R. Prakken [1898–1978]	M. Gandhi [1869–1948]
↓	↓	
E. M. East [1897–1938]	Jan Antonie Honing [1880–1950]	
↓	↓	
C. G. Hopkins [1866–1920]	H. de Vries [1848–1935]	
↓		
B. Tollens [1841–1918]		
↓		
F. Wöhler [1800–1882]		

Legend for M. S. Swaminathan Pedigree

Mankombu Sambasivan Swaminathan (b. 1925) was born in Kumbakonam, India. He is the father of the Green Revolution in India. He produced high yield wheat after learning genetics at Wisconsin in 1952. He was motivated by the great famine in India which he experienced in his youth.

Roelof Prakken (1898–1978) was born in Enten and died in Wageningen, both in Netherlands. He obtained his PhD at the University of Utrecht. He studied quantitative traits in beans.

Jan Antonie Honing (1880–1950) was born in Amsterdam and died in Wageningen, both in Netherlands. He studied wheat cultivars. His PhD was in **de Vries's** laboratory studying Oenothera.

Hugo de Vries (1848–1935) was born in Haarlem and died in Lunteren, both in Netherlands. He studied plant physiology and contributed to the role of osmosis in plant cells before he shifted to genetics and was a re-discoverer of Mendel's laws. He proposed as theory of intracellular pangenesis in 1888 (genes did not circulate from cell to cell) and he introduced the concept of mutation as a genetic event involving one or more genes. His mutations leading to new species of Oenothera turned out to be polyploids, aneuploids, and chromosomal rearrangements rather than Darwinian gene mutations, the more common differentiation among cognate species.

Royal Alexander Brink (1897–1984) was born in Woodstock, Ontario, Canada, and died in Madison, Wisconsin. He became a leader in maize genetics, working on transposal elements, pollen developmental genes, and the genetic role of the endosperm.

Edward Murray East (1879–1938) was born in Du Quoin, Illinois, and died in Boston, Massachusetts. He developed hybrid corn as a commercial success for United States farmers. He studied inbreeding and wrote widely on that topic. He was also an advocate of the negative eugenics movement, favoring compulsory sterilization of socially unfit people and restrictive immigration based on ethnic bigotry.

Cyril George Hopkins (1866–1920) was born in Chatfield, Minnesota, and died in Gibraltar. He was an agronomist who developed methods of crop rotation to provide soil fertility. He went to Greece to help their agriculture and contracted malaria in Gibraltar and died just before heading back to the United States.

Bernhard Tollens (1841–1918) was born in Hamburg and died in Göttingen, both in Germany. He was a chemist and worked on carbohydrates, identifying the sugars that composed them.

Friedrich Wöhler (1800–1882) was born Frankfurt and died in Göttingen, both in Germany. He was a founder of the field of organic chemistry, synthesizing molecules once thought only plants or animals could make. His production of urea from non-living chemicals was one of many findings in this new field.

Mahatma Gandhi (1869–1948) was born in Porbandar and died in New Delhi, both in India. He became a lawyer and led the non-violent civil disobedience movement after reading about **Henry David Thoreau's** career. Gandhi led a boycott of imported fabric and clothing from Europe. He gained independence for India but was assassinated in 1948.

C. C. Tan [1909–2008]

↙   ↓   ↘

Ju-Chi Li [1895–1991]   Zin-Ying Chen [b. 1897]   T. Dobzhansky [1900–1975]

↓   ↙   ↓

T. H. Morgan [1856–1945]   ←   A. H. Sturtevant [1891–1970]

↘

See pedigree of **H. J. Muller**

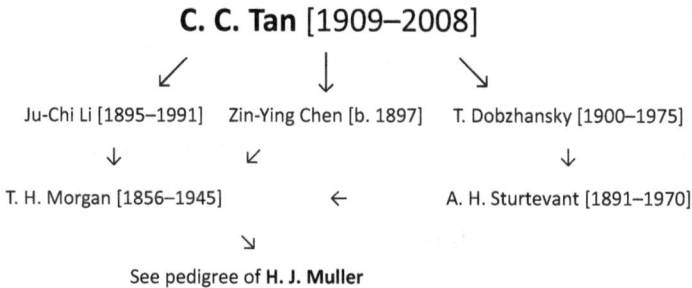

Legend for Pedigree of C. C. Tan

C. C. Tan [**Tan Jiazhen**] (1909–2008) was born in Zhejiang, China, and died in Shanghai, China. He studied with **Li** and **Chen** for a MA degree and went to **Morgan**'s fly lab in California for his PhD which he got with **Dobzhansky**. He studied chromosomal breakage and the evolution of the Drosophila genus. He produced a phylogenetic tree showing the historical progression. He returned to China and faced hardships during the Lysenko era and equal hardship during the Cultural Revolution. After these had passed, he was rehabilitated and made rector of Fudan University. His property was restored. He lived to be 99.

Ju-Chi Li (1895–1991) was born in Tientsin, China. He went to the United States for his education, doing undergraduate work at Purdue (where he played football) and then went to Caltech to work with **Morgan**'s fly lab, getting his PhD in 1936. He studied developmental effects of mutations on *D. melanogaster.*

Zin-Ying Chen (b. 1897) was born in China and worked in **Morgan**'s laboratory, getting his PhD in 1926.

Note: Morgan had four students from China. In addition to **C. C. Tan, Ju-Chi Li**, and **Zin-Ying Chen**, he had **Shisan C. Chen** (1894–1957) who got his MA in 1921 at Columbia. S. C. Chen devoted his career, on his return to China, to the genetics of gold-fish, tracing the history of its variants back to 1000 years of cultivation. He showed their Mendelian nature.

Nikolai Vavilov [1887–1943]

↙   ↓

William Bateson [1861–1926]   Rowland Biffen [1874–1949]

↙   ↘   ↓

F. M. Balfour   William K. Brooks   Harry M. Ward
[1851–1882]   [1848–1908]   [1854–1906]

↓   ↓   ↙   ↘

Michael Foster   →   See **Muller** pedigree   Thomas H. Huxley   W. T. Thiselton-Dyer
[1836–1907]   to **Galileo**   [1825–1895]   [1843–1928]

↓

See **Muller** pedigree to **Newton**

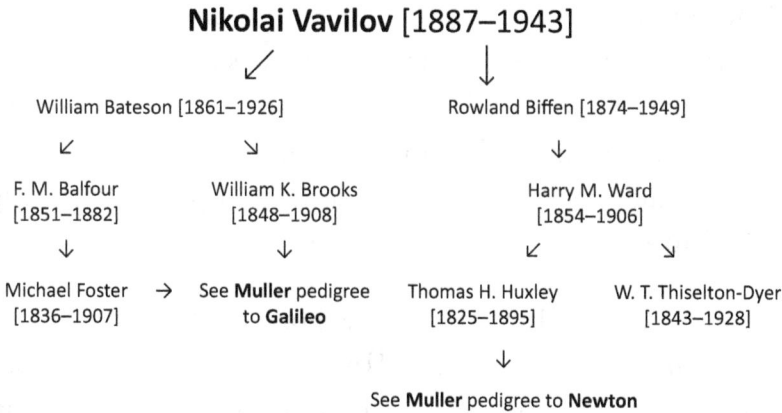

Legend for Pedigree of Nikolai Vavilov

Nikolai Vavilov (1887–1943) was born in Moscow and died in Saratov, both in Russia. He studied agriculture in 1910 and in 1913 went to England to study with **Bateson** and **Biffen**. He took an interest in both seed production in relation to their environments and in the centers of origin where domesticated plants came from. This took him on many trips around the world. He created a huge seed bank in St. Petersburg, and it survived the siege of Leningrad. Vavilov lost his job and was arrested by supporters of Lysenkoism in 1939 and sent to prison in Siberia where he died.

Rowland Biffen (1874–1949) was born in Cheltenham, England, and died in Cambridge, England. He taught agricultural botany and demonstrated Mendelism in wheat. He also analyzed wheat varieties and showed their varieties were Mendelian in the genetic ratios.

Harry Marshall Ward (1854–1906) was born in Hereford, England, and died in Babbacombe, England. He taught agricultural science and showed how fungal infections could be prevented or reduced by strip planting areas for growing coffee and boundary areas with leafy plants that prevent wind-blown spores from getting to the coffee plants. He also believed in keeping numerous varieties of domesticated plants to prevent monoculture which is vulnerable to massive crop failure.

William Bateson (1861–1926) was born in Whitby, England, and died in Merton, England. His father was a college master. Bateson took an interest in embryology from courses with **Weldon** and **Balfour** and did postgraduate work with **Brooks** at Johns Hopkins. He studied *Balanoglossus* and showed its relation to the evolution of the vertebrates. Brooks recommended he take on heredity as the major area of biology needing a fresh approach. Bateson studied variations and came up with homeotic and meristic mutations which he associated with major shifts of organ location and number. He confirmed Mendelism and discovered epistatic relations of genes leading to aberrant ratios. He coined the term genetics.

William T. Thiselton-Dyer (1843–1928) was born in London and died in Gloucestershire, England. His father was a physician. He chose botany as a career and traveled to the British colonies, choosing plants to cultivate in Kew and other gardens. He also did experiments to find plants in one British colony that would grow well in another British colony (e.g., cacao in Ceylon).

William Keith Brooks (1848–1908) was born in Cleveland and died in Baltimore. He had frail health from a congenital heart defect. He was an avid scholar and read widely in classics and philosophy before committing himself to natural history after reading

Darwin's *Origin of Species*. At Harvard, he got his PhD with Louis and Alexander Agassiz. He studied embryology of tunicates (the genus *Salpa*) and invertebrates, especially mollusks. He believed a study of tunicates would contribute to the understanding of the evolution of vertebrates from their invertebrate ancestors. He was a gifted teacher and among his PhD and post-doctoral students were William Bateson, T. H. Morgan, Edmund B. Wilson, H. V. Wilson, and Ross Harrison. He wrote seven influential books, including *The Law of Heredity* (1883). His views were Lamarckian. He told Bateson that the most interesting area he could enter was heredity. He instructed his students to work with living specimens and study their living functions. Bateson followed that advice but rejected Brook's theoretical approach.

Thomas Henry Huxley (1825–1895) was born in Ealing, England and died in Eastbourne, England. His father was a mathematics teacher and when the school failed, young Huxley had to drop out of his father's school at the age of 10. He became an autodidact and taught himself Latin, Greek, and German and read widely. He served as an apprentice to surgeons and joined the Navy as an assistant surgeon. He was sent on the *HMS Rattlesnake* to an expedition to New Guinea and Australia. From his dredging at sea, he collected numerous medusae and classified them. His surgical skills were exceptional, and he found these organisms had only two layers instead of three (they lacked mesoderm). His papers on what he called the Hydrozoa (today called Cnidarians) earned him election to the Royal Society at the age of 26. Huxley became friends with Charles Darwin and while he was initially skeptical of evolution and natural selection (which Darwin confided to him), he was won over by the evidence and became a strong supporter when Darwin's theory cam out. Huxley taught at the London

School of Mines and there he developed a theory of liberal arts education which was influential in Europe and North America. His essay "A liberal education and how to get it" was published in 1868. His essay "On a piece of chalk" is one of the most captivating public lectures ever given on the significance of evolution. Huxley rejected formal religious belief and coined the term agnostic to represent his views that if a belief lacks scientific demonstration it can neither be proven nor disproven. He major students were Henry Newell Martin and Michael Foster.

Conrad Waddington [1905–1975]

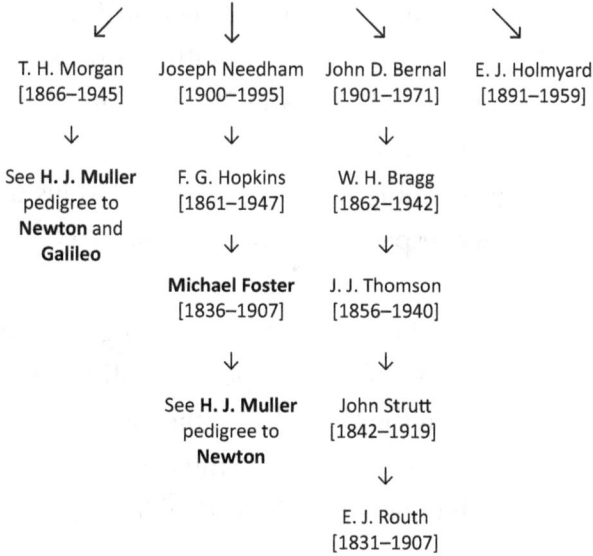

↙	↓	↘	↘
T. H. Morgan [1866–1945]	Joseph Needham [1900–1995]	John D. Bernal [1901–1971]	E. J. Holmyard [1891–1959]
↓	↓	↓	
See **H. J. Muller** pedigree to **Newton** and **Galileo**	F. G. Hopkins [1861–1947]	W. H. Bragg [1862–1942]	
	↓	↓	
	Michael Foster [1836–1907]	J. J. Thomson [1856–1940]	
	↓	↓	
	See **H. J. Muller** pedigree to **Newton**	John Strutt [1842–1919]	
		↓	
		E. J. Routh [1831–1907]	

Legend for Academic Pedigree of Conrad Waddington

Conrad Hal Waddington (1905–1975) was born in Eresham, England, and died in Edinburgh, Scotland. His parents moved to India when young Waddington was 3. The following year they sent Waddington to his uncle and aunt's home back in England. He enjoyed his secondary school mentor, **Eric John Holmyard,** who taught him philosophy and introduced him to Eastern mysticism. In WWII Waddington served in the Royal Air Force. He had studied embryology for his PhD and was mentored by **Needham** and **Bernal**. He also shared their communist political views. He took a holistic view of developmental biology, not in the sense of vitalism, but in a model of interactions that he called genetic assimilation, epigenetic landscapes, and canalization. He also learned fruit fly genetics by going to California to work a year with **Morgan's** fly lab on wing variations.

Noel Joseph Terence Montgomery (Joseph) Needham (1900–1995) was born in London and died in Cambridge, both in England. He was an only child; his father a physician and his mother, a composer. He got his BA in 1921, MS in 1925, and PhD in 1925 at Cambridge University. By 1931 he wrote a three-volume overview of chemical embryology. He then shifted his interests to the history of science, especially China and learned Chinese so he could read original documents which he collected from Chinese scholars. He showed that China invented printing, gunpowder, and many other technologies before European scientists. He attributed the success of science in the West due to the diversity of European countries and institutions in contract to the Emperor-oriented society in China which did not exploit individual findings.

Frederick Gowland Hopkins (1861–1947) was born in Eastbourne, England, and died in Cambridge, England. He obtained his MD at King's College, London. He was an early contributor to biochemical studies of cell functions and related oxygen depletion to muscle production of lactic acid. He showed that margarine lacked vitamins A and D and recommended they be added to promote health of those not using butter. He brought biochemistry into the medical school curriculum.

John Desmond Bernal (1901–1971) was born in Nenagh, Ireland, and died in London, England. He was of Sephardic Jewish ancestry on his father's side and raised as a Catholic which he abandoned later as an atheist. His mother was a journalist. He studied X-ray crystallography with **William Bragg**. He shifted to biological molecules, working the structure of cholesterol and steroids derived from it. He studied tobacco mosaic virus and showed its structural shape.

William Henry Bragg (1862–1942) was born in Wigton, England, and died in London, England. With his son **Lawrence Bragg** they applied X-rays to study crystal structure for which they received a Nobel Prize.

Joseph John Thomson (1856–1940) was born in Manchester, England, and died in Cambridge, England. He received a BA in mathematics in 1880 and switched to physics. His work on the passage of electric current in gases led him to infer the existence of an atomic particle with negative charge, the electron, which had 1/1000th the mass of a hydrogen atom. It led to the field of atomic particle physics. His tentative model of the atom was likened to a pudding in which one or more electrons were embedded.

John Strutt (Baron Rayleigh) (1842–1919) was born in Maldon and died in Witham, both in England. He studied the density of gases and discovered the element argon. He also showed how light is scattered by small particles present in gas and why this makes the sky look blue.

Edward John Routh (1831–1907) was born in Quebec, Canada, and died in Cambridge, England. He was a coach for the Mathematical Tripos, the lead student being a senior wrangler. He coached over 600 students and 28 became senior wranglers which is a record that still stands.

Eric J. Holmyard (1891–1959) was a science teacher at Clifton College with interests in the history of science and Arabic science. His mentoring helped shape **Waddington's** interdisciplinary and holistic approach to science.

James Dewey Watson [b. 1928]

↙   ↓   ↘

Salvador Luria [1912–1991]   Max Delbrück [1906–1981]   H. J. Muller [1890–1967]

↓   ↓   ↓

See tree   See tree   See tree

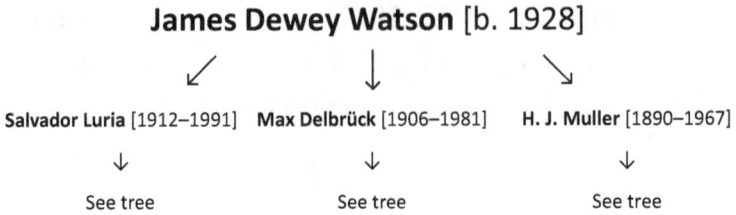

Legend for J. D. Watson Pedigree

James Dewey Watson (b. 1928) was born and raised in Chicago, Illinois. His memoir, *The Double Helix* (1968), became a bestseller. His initial interest was in ornithology, but **Fernandus Payne** told Watson that if he wanted to go into that field, he should apply to Cornell University. If he wanted first rate genetics, he should come to Indiana University. Payne was a student of **Morgan** and **Wilson** and active in the IU Zoology Department when Watson came to Indiana University for his PhD. At the time, Indiana University had four world class geneticists — **Muller** using fruit flies, **Sonneborn** using Paramecia, **Luria** using viruses, and **Cleland** using Oenothera. Watson chose **Luria** for his dissertation work because he felt the study of virus genetics was most likely to reveal gene action and composition. He also felt **Muller's** courses stressing the gene were the most influential in shaping his career. His work with **Francis Crick** at Cambridge led to their 1953 papers in *Nature* that provided the chemical structure of DNA and the biological implications of that structure for gene replication, mutation, and function. Through Luria, Watson became a member of the phage group that **Delbrück** founded when he went to Caltech after leaving Germany and settling into the United States. Delbrück's quest for the physical structure of the gene resonated with Watson. Watson also helped launch the Human Genome Project. Watson's personality was controversial because he spoke his mind on issues with a tart tongue causing responses in kind. He believed that human intelligence was tied to heredity and this made him a racist in the interpretation of some of his critics. He said his son was schizophrenic and he loved his son but recognized genetic components contributed to that condition. He was accused of being a sexist in his treatment of **Rosalind Franklin** in his memoir, but he said he was reflecting how most males of his generation reacted to females but never doubted her competence as a scientist. He denied that seeing her photo of DNA enabled him to figure out the structure of DNA. He said it confirmed what he and Crick were already considering — a double helix for DNA.

Author's note: I much appreciate Watson's help in my career. It was Watson who praised my biography of Muller that boosted my career in the history of science. He also read my draft of *The Unfit: A History of a Bad Idea* and recommended CSHL Press to publish it. He said it was important for CSHL Press to publish books critical of the American eugenics movement because we should not repeat mistakes of the past. It is ironic that he was in England promoting one of his books and visiting "girl's schools" to recruit females for PhDs in the Watson School at CSHL while his critics were calling him sexist for his portrayal of Rosalind Franklin.

August Weismann [1834–1914]
↓

Rudolf Leuckart [1822–1898]
↓

Rudolf Wagner [1805–1864]

↙   ↘

Johann Schönlein [1793–1864]   Georges Cuvier [1769–1832]

↙   ↙   ↓

Friedrich Tiedemann [1781–1861]   Abraham Werner [1749–1817]   ↓

↓   ↓   ↓

Johann W. C. Bruhl [1757–1806]   Johann Gehler [1732–1796]   Henri Tessier [1741–1837]

↓

Christian Ludwig [1709–1773]

Legend for Intellectual Pedigree of August Weismann

August Weismann (1834–1914) was born in Frankfurt and died in Freiburg, both in Germany. His father taught classics and his mother played the violin. He got his MD in 1859 at Göttingen. He began losing his vision in 1864. This limited periods when he could work with a microscope. He studied insect metamorphosis, parthenogenesis, gamete formation, and variations in butterfly wings. He wrote many theoretical books based on these studies. He introduced the germplasm theory based on the setting aside of germ cells in the early embryo of insects. He argued that somatic cells could not pass on genetic specificities to the germ line by his experimentation on mutilated mice and a search of the literature (circumcision and head binding were not transmitted). He championed Darwin and natural selection.

Rudolf Leuckart (1822–1898) was born in Helmstedt and died in Leipzig, both in Germany. He was a parasitologist studying tapeworm infection and trichinosis in pork. He broke in two **Cuvier**'s Radiata and showed they consisted of two separate phyla, coelenterates (cnidarians) and echinoderms. He discovered the micropyle and its relation to insemination in insects. He also studied parthenogenesis.

Rudolf Wagner (1805–1864) was born in Bayreuth and died in Göttingen, both in Germany. His father was a professor. Wagner got his PhD in 1826 at Erlangen and studied the anatomy of nerve cells, comparative anatomy, and described the germinal vesicle. He rejected materialism and favored **Goethe's** naturphilosophie.

Johann Schönlein (1793–1864) was born and died in in Bamberg, Germany. He got his MD and studied the cause of ringworm (*Trichophyton schoenleinii*). He was one of the first German professors to lecture in German instead of Latin. He gave the name tuberculosis to what was called consumption. He was the first physician to introduce mycology (the study of fungi) in the medical curriculum.

Friedrich Tiedemann (1781–1861) was born in Kassel and died in Munich, both in Germany. He obtained his MD and studied comparative anatomy, the fish heart, amphibian anatomy, and the orangutan anatomy. He also studied the embryology of human organ systems from studies of miscarried fetuses or dead pregnant women. He compared brains and body size and showed no differences in brain anatomy of male and female whites or blacks. He discredited what was called "scientific racism." He was an abolitionist. He also worked on digestion and showed that digestive juices were involved in breaking down starchy foods into sugars.

Georges Cuvier (1769–1832) was born in Montpelier and died in Paris, both in France. He was a Lutheran and his father was in the Swiss Guard. He was a founder of the fields of comparative anatomy and paleontology. He believed extinctions were real and he named several extinct forms (mastodons, pterodactyls) and believed there were several massive floods that shifted the earth from one geological era to another. He rejected **Lamarck's** theory of evolution of acquired characteristics and favored a series of new creations as each era died out, usually by a catastrophe.

Henri Tessier (1741–1837) was born in Angerville and died in Paris, both in France. His father was a notary. He studied agricultural science. He collected wheat varieties and tested those most suited for French farmers. He cultivated sugar beets and taught how sugar could be extracted from them. He imported merino sheep and selected varieties for their wool properties.

Abraham Werner (1750–1817) was born in Wachau, Slovenia, and died in Dresden, Germany. His father was a foreman in a foundry. He enjoyed research in mineralogy and wrote the first text in that field. He knew mining engineering and believed that a series of rock layers were formed because of rising or falling water levels around the earth. He called this interpretation Neptunism. He classified these layers as igneous, sedimentary (limestone), alluvial (sands, clay, gravel), mountainous (caused by pressure uplifting land) and volcanic (pumice and lava). For this reason, he is regarded as the founder of stratigraphy in geology.

Johann Carl Gehler (1732–1796) was born in Göritz, Germany, and died in Leipzig, Germany. His father was a miner. He taught minerology and then shifted to obstetrics, teaching midwifery and writing a book on obstetrics.

Christian Ludwig (1709–1773) was born in Brzeg, Poland, and died in Leipzig, Germany. He got his MD at Leipzig and studied intestinal worms, sex in plants, and corresponded with **Linnaeus** on numerous plants, some of which were new species.

Jan Anthony Witkowski [b. 1947]

↙   ↓

William Brighton   Victor Dubowitz [b. 1931]

↙   ↘

Ronald Illingworth [1909–1990]   Everson Pearse [1916–2003]

↓

Arnold Gesell [1880–1961]

↓

Joseph Jastrow [1863–1944]

↓

Charles Sanders Peirce [1839–1914]

↓

William James [1842–1910]

↓

Louis Agassiz [1807–1873]

↓

See **Muller** pedigree for **Agassiz** to **Galileo**

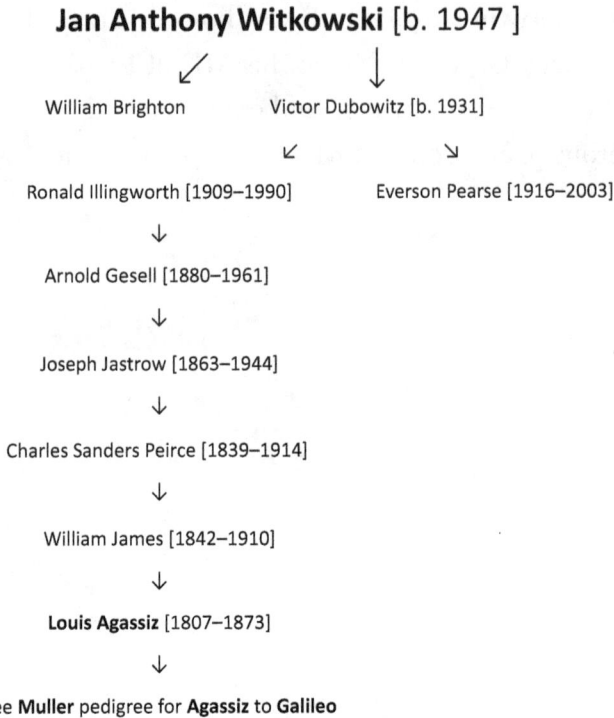

Legend for Intellectual Pedigree of Jan Anthony Witkowski

Jan Anthony Witkowski (b. 1947) was born and raised in Dimmingham England and studied at the University of Southampton for his BSc. He studied muscular dystrophy and muscle physiology at the University of London for his PhD. He developed interests in the history of life sciences, especially developmental biology and became Director of the Banbury Center Conferences at Cold Spring Harbor. He edited many books and journals and he joined the Watson School at Cold Spring Harbor Laboratory where he has taught since its inception.

Victor Dubowitz (b. 1931) was born in Beaufort East, S. Africa and became a neurologist.

Ronald Illingworth (1909–1990) was born in Harrogate, England and died in Bergen, Norway.

Arnold Gesell (1880–1961) was born in Alma, Wisconsin. He was a clinical psychology specializing in child development. He received his MD at Yale.

Joseph Jastrow (1863–1944) was born in Warsaw, Poland. He was a psychologist who specialized in optical illusions.

Charles Sanders Peirce (1839–1914) was born in Cambridge, Massachusetts, and died in Milford, Connecticut. He was a psychologist and philosopher and a founder of American Pragmatism. He suffered from facial nerve neuralgia (tic dolorous) and this made academic employment impossible for him to sustain.

William James (1842–1910) was born in NYC and died in Chocorua, New Hampshire. He was the co-founder of pragmatism with Peirce and is considered the father of American psychology. He was a participant in the Metaphysical Club in Boston. He also went with **Louis Agassiz**, whom he knew at Harvard, and joined him for an eight months trip down the Amazon River in Brazil. James was largely an autodidact. He said, "The first lecture on psychology that I ever heard was the first I ever gave."

Sewall Wright [1889–1988]
↓

William E. Castle [1867–1962]

↙   ↓

Clarence L. Herrick [1858–1904]   Edward L. Mark [1847–1946]

↘   ↓

Rudolf Leuckart [1822–1898]

↓

Rudolf Wagner [1805–1864]

↓

Johann Schönlein [1793–1864]

↓

Ignaz Döllinger [1770–1841]

↓

See **H. J. Muller** pedigree **to Galileo**

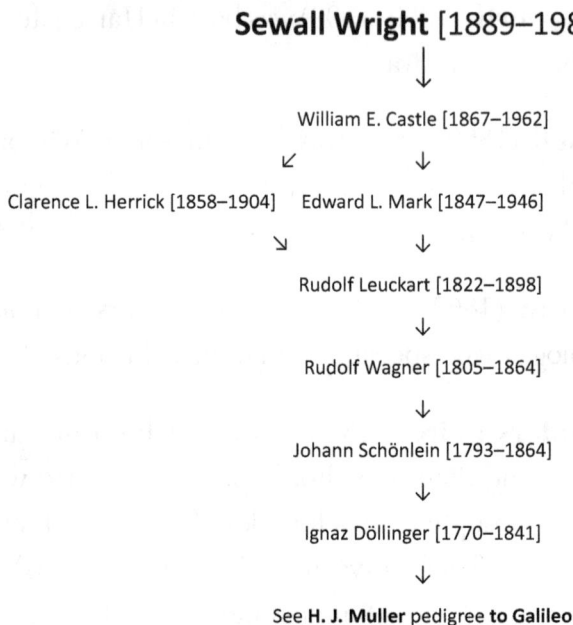

Legend for Sewall Wright Pedigree

Sewall Wright (1889–1988) was born in Melrose, Massachusetts, and died in Madison, Wisconsin. He obtained his PhD with William Castle studying coat color in mammals. He shifted to population genetics and introduced mathematical models for founder effects, drift, and mutation influence on population diversity and natural selection.

William Ernest Castle (1867–1962) was born in Alexandria, Ohio, and died in Berkeley, California. He received his PhD at Harvard and studied embryology and zoology. He was one of the first American scientists to shift to Mendelian heredity and promote Mendelism, using small mammals and coat color for confirmation. He studied variable traits and their genetic mechanisms.

Clarence L. Herrick (1858–1904) was one of the founders of ethology, studying animal behavior, especially birds, and its relation to evolution.

Edward Laurens Mark (1847–1946) was born in Hamlet, New York, and died in Cambridge, Massachusetts. He received a BA in 1871 from the University of Michigan and a PhD in 1876 in Leipzig, Germany, with **Leuckart** as his mentor. Among his more famed students at Harvard were **Davenport, Castle,** and **Jennings**.

Karl Rudolf Leuckart (1822–1898) was born in Helmstedt, Germany and died in Leipzig, Germany. He became a parasitologist, working out the life cycles and hosts for trichinosis, human and animal tapeworms, and liver flukes. His work provided the basis for meat inspections in Germany and throughout the industrialized world.

Rudolf Wagner (1805–1864) was born in Bayreuth, Germany, and died in Göttingen, Germany. He interpreted the germinal vesicle on the ovary as a container of the ovum which is arrested in development at Prophase I of meiosis until a surge of luteinizing hormone resumes development in Meiosis II. He was also a vitalist who followed **Goethe's** idea of nature philosophy.

Johann Schönlein (1793–1864) was born and died in Bamberg, Germany. He was the first German professor to teach his classes in German instead of Latin. He also identified ringworm as a parasitic infection of a mold, *Trichophyton schönlandeii*. He also changed the name of consumption to tuberculosis because he identified the microscopic tubercles in the lungs.

Ignaz Döllinger (1770–1841) was born in Bamberg and died in Munich, both in Germany. He went to Italy to study biology, obtaining his doctorate with **Antonio Scarpa** at Padua. He studied embryology using microscopy upon his return to Germany. Among his students were **Agassiz, von Baer, Pander, Oken,** and **Schönlein,** who spread Italian descriptive science and German laboratory science across Europe.

Note: See Döllinger to Galileo in Muller reference pedigree.

Afterword

I believe we can infer some generalizations from a study of these intellectual pedigrees. The route from Medieval (ca 1200–1400s) to Renaissance (1500–1600s) to Enlightenment (1700s), to early Modern (1800s), and to Modern (1900-present) Biological Science is roughly from Italy to Germany, to France, to England and the United States. It is **Döllinger** who goes to Italy to work with **Scarpa** to bring Italian science to Germany. There it blossoms into the German University model of the PhD through the **Humboldt** brothers. It spreads to France as **Cuvier** develops comparative anatomy and **Lamarck** offers the first serious efforts to use evolution to explain life on earth. **William Harvey** in England is also impressed by Italian medical science which he learned at Padua (while **Galileo** was on the faculty) and brings that physiological approach to a successful interpretation of the circulation of the blood. **Darwin** is strongly influenced by French evolutionary theory but keeps an open mind as he travels around the world to study life's diversity and the reasons for its present state. The New World is the beneficiary of importing Swiss naturalist **Agassiz** from Paris to establish Harvard as the first American University to stress the importance of the PhD as the entry to university professorship. At Johns Hopkins University, **Gilman** travels to Europe recruiting faculty to stress experimental approaches to science. He brings **Martin** (a **Huxley** product) and hires Brooks (a student of **Agassiz**) to mentor American graduate students.

When we think of science today at the university level we think of the major fields — physics, astronomy, chemistry, biology, geology, and psychology as the major experimental sciences. The life sciences are often subdivided into smaller departments — zoology, botany, anatomy, physiology, genetics, cell biology, biochemistry, microbiology, evolution, ecology, and many others. Those departments in most universities have a department head and 2 or more faculty. That was not the case before the start of the twentieth century. Many of these fields did not exist and a more traditional early nineteenth century university had botany and zoology in the Arts and Sciences faculty and anatomy and physiology in the medical school faculty. Almost all faculty in the life sciences before the 1800s had the MD as their objective. In Medieval and Renaissance times botany was part of the medical curriculum because almost all pharmaceutical products were derived from plants. When I taught at Queen's University in Kingston, Ontario (1958–1960), I was surprised that the head of the Biology Department, **Gleb Krotkov**, was not allowed to hire a biochemist because "biochemistry belongs to the medical faculty."

Not all life scientists were educated in a medical curriculum. **Robert Hooke** and **Antony van Leeuwenhoek** were not MDs. Hooke was a physicist and **Leeuwenhoek** was a linen draper, a rare instance of an autodidact playing a major role in science. Several had difficult childhoods (infirmity or poverty). **Tartaglia** was deformed by cuts from a saber as an infant making speech difficult. **Berzelius**'s father died when young Berzelius was 4 and his mother died when he was 10. He was raised by his alcoholic aunt. **Pauling**'s father died when young Pauling was 10 and he supported his education taking odd jobs from high school through college. Some life scientists had home-schooled education or were autodidacts who never went to college. This includes **Linnaeus, Herbert Spencer**, and chemist **Robert Boyle**. Some were women who were barred from attending college. **Florence Nightingale**, founder of the Red Cross and the professionalization of nursing, put together her course syllabus by visiting nursing services in hospitals in the US and Europe.

Medieval manuscripts showing illustrations of the Master teaching students show about ten students, usually with the Master commenting on readings from a book (usually Roman or Greek scholarly works). The oldest university, at Bologna in Italy, had several noted alumni: **Dante Alighieri, Erasmus of Rotterdam** (author of *In Praise of Folly*), **Marcello Malpighi, Albrecht Dürer, Nicolas Copernicus, Paracelsus, Lazzaro Spallanzani,** and **Luigi Galvani.**

Besides the university, the introduction of scientific academies played a role in spreading science in the Renaissance. The first Academy of science was established in Naples in 1560. In 1603, The Academy dei Lincei was established in Rome (**Galileo** was a member). Florence formed its academy in 1657. The Royal Society met informally from 1648–1660 and then became chartered by **Charles II** in 1652 as a national (royal) academy. Similarly, the French scientists met informally at the royal library and then moved to the Louvre as an Academy of Sciences in 1699. In the US, **Abraham Lincoln** signed a bill passed by Congress in 1863 establishing the National Academy of Sciences. It had 50 founding members including **Louis Agassiz** and naturalist and evolutionist **Asa Gray**. Also, in the US the National Research Council was established by **President Wilson in** 1916, the National Academy of Engineering in 1964, and the National Institutes of Medicine in 1970 (which is now, since 2015, the National Academy of Medicine). Of these the most famous is the Royal Society of Science because its publications were internationally circulated, and it accepted contributions for peer review from all countries. **Leeuwenhoek** and **Malpighi** published frequently in the *Proceedings of the Royal Society*.

There are many aspects of the history and philosophy of science which are either not known or not well known. These include networks of scientists who share interests in an emerging field of science and how that network forms. To outsiders these look like "in groups" and "outgroups" as if they were exclusionary by intent. When **Leslie Dunn** hoped to work with **Morgan** at Columbia, about 1914, Morgan told him his laboratory was full and Dunn went to Harvard and had **Castle** as his sponsor. Dunn did not feel that he was being snubbed. He felt his timing was off. Sometimes the "in group" reputation is deserved. **Edgar Altenburg** told me, when I interviewed him for the Muller biography that he asked Morgan if he could join the fly lab. Morgan dipped his finger in an aquarium tank, held a drop of water to a lamp bulb, and said, "There are a lot of Daphnia in this drop, why don't you study them?" He took it as a put-down and Muller served as his conduit to the work going on in the fly lab.

When I wrote *The Gene: A Critical History* (1966) I was struck by the personality differences among geneticists whose published papers I read at the Marine Biology Laboratory at Woods Hole. Before 1920 there were gratuitous remarks in the publications of **Muller** and **Castle**, Muller and **Goldschmidt**, Goldschmidt and **Sturtevant**, and other geneticists. The best studied case is that of **Bateson** in his disputes with **Karl Pearson** and **W. F. R. Weldon**. After WW1 these were kept out of the publications. I assume (but have not yet found) that there was a meeting or correspondence among editors of a few key journals that agreed to edit these out of accepted manuscripts and keep the debates on the merits of the evidence presented and not the implied bias of the author. The animosity goes back to the dawn of modern science. **Galileo**'s books used satire to characterize the views of his opponents. **Cuvier**'s obituary of **Lamarck** is devastating in its character assassination. In modern times, **J. D. Watson**'s memoir on *The Double Helix*

(1968), caused an uproar of criticism for his portrayals of **Francis Crick**, **Rosalind Franklin**, and other participants or competitors in X-ray crystallography and biochemical or molecular biology. While I favor an editorial policy not to introduce polemics into scientific articles, I feel it is the job of historians of science and philosophers of science to explore these personality traits. They do go back to the earliest writings on science (**Plato** certainly used his Socratic dialogues to put down his critics), but how extensive or necessary are such strategies to launch new scientific findings, tools, or theories? **Sigmund Freud** tried to interpret **Leonardo da Vinci**'s psyche using his "association" approach and using recorded dreams, anecdotes, diaries, correspondence and other informal writings to locate such evidence for personality traits.

I hope readers of this book will construct their pedigrees with this advice in mind:

1. Start with your name or your mentor's name using a Google search.
2. First get the sequence down as far as you can follow it. Wikipedia entries are your primary source for information.
3. After you have the names and dates of birth and death get the places of birth and death.
4. Go back to see who the parents were and what they did (about a third of the 65 entries have some sort of information on upbringing, especially the father's career).
5. Take notes on each entry, what their education was like and where they got it.
6. Add the most important finding or findings of each scientist in your pedigree.
7. Add personal information (cause of death), marital status, personality, military service, illness, bizarre behavior.

I hope there will be a historian of science who starts a repository for academic pedigrees accessible online. It will be a wonderful reference for those future scientists preparing their own pedigrees. It will also be a rich source of data for those in the history, philosophy, and sociology of science to see trends and connections and a more enriched view of how fields emerge and develop.

Intellectual pedigrees serve several purposes. They give a history of science over several centuries as do "timelines" for history (first introduced by chemist and Unitarian theologian **Joseph Priestley**). They are thus an educational tool. They provide a sense of "belonging" just as family pedigrees do. I feel inspired looking at my name attached to Muller's academic pedigree. They are thus personalized histories that could be characterized as feeding one's vanity (as do personal genealogies for one's family). But just as family pedigrees can reveal psychosis, alcoholism, felonies, children born out of wedlock, ancestors who entered a country illegally, and personal triumphs and tragedies, so too do these intellectual pedigrees when the biographies are looked up and selected episodes are used to convey an intellectual ancestor's life. They can inspire pride, inspiration, or imitation especially if students are looking at a framed intellectual pedigree in a scientist's office or home. It is my hope that a comparative study of intellectual pedigrees in different fields of science will reveal similarities and differences in how mentoring works and how fields of science emerge and evolve.

References and Recommended Reading

Carlson Elof (1966) *The Gene: A Critical History*. W. B. Saunders Publishers, Philadelphia.

Carlson Elof Axel (2017) Speaking out about the social implications of science: The uneven legacy of H. J. Muller. *Genetics* 187: 1–7. https://doi.org/10.1534/genetics.110.125773

Carlson Elof Axel (2017) Scientific feuds, polemics, and ad hominem arguments in basic and special-interest genetics. *Mutation Research* 771: 128–133. https://doi.org/10.1016/j.mrrev.2017.01.003

Harmon Oren and Michael R. Dietrich, editors (2018) *Dreamers, Visionaries, and Revolutionaries in the Life Sciences*. University of Chicago Press, Chicago.

Sources of Illustrations

Figure Legend or Name

1 Galton pedigree from Laughlin Archives, Missouri State University
2 A. H. Sturtevant 1965 *A History of Genetics* Harper and Row, New York p. 140
3 Elof Carlson, personal photo
4 Elof Carlson, personal photo
5 Morris Cohen, personal photo
6 H. J. Muller, personal photo
7 T. H. Morgan, Wikipedia public domain
8 H. N. Martin, Wikipedia public domain
9 M. Foster, Wikipedia public domain
10 T. H. Huxley, Wikipedia public domain
11 C. Darwin, Wikipedia public domain
12 A. Sedgwick, Wikipedia public domain
13 T. Jones, Wikipedia public domain
14 T. Postlethwaite, Wikipedia public domain
15 S. Whissom, Wikipedia public domain
16 R. Smith, Wikipedia public domain
17 R. Cotes, Wikipedia public domain
18 I. Newton, Wikipedia public domain
19 E. B. Wilson, Wikipedia public domain
20 W. K. Brooks, Wikipedia public domain
21 L. Agassiz, Wikipedia public domain
22 G. Cuvier, Wikipedia public domain
23 I. Döllinger, Wikipedia public domain
24 A. Scarpa, Wikipedia public domain
25 G. Morgagni, Wikipedia public domain
26 A. Valsalva, Wikipedia public domain
27 G. Malpighi, Wikipedia public domain

About the Author

Elof Axel Carlson is a geneticist, historian of science, and writer. He attended NYU as an undergraduate with a BA majoring in biology and minoring in history. He received his PhD in 1958 at Indiana University studying with Nobelist, Hermann J. Muller. Carlson has taught at Queen's University in Kingston, Ontario, at UCLA, and at Stony Brook University where he retired as emeritus Distinguished Teaching Professor in the Department of Biochemistry and Cell Biology. He lives since 2009 in Bloomington, Indiana, and is a Visiting Scholar in the IU Institute for Advanced Study. He is the author of 14 books including *The Gene: A Critical*

History (1966); *Genes, Radiation, and Society: The Life and Work of H. J. Muller* (1982); *The Unfit: A History of a Bad Idea* (2001), *Mendel's Legacy: The Origin of Classical Genetics* (2004), and *The 7 Sexes: Biology of Sex Determination* (2012). His most recent book was *How Scientific Progress Occurs: Incrementalism and the Life Sciences* (2018). He has also written a newspaper column, *Life Lines*, for the North Shore of Long Island newspapers since 1997. Carlson is married to Nedra (née Miller) a retired clinical IVF embryologist. They raised five children and have 12 grandchildren and three great-grandchildren.

Index